Marcus Gutjahr

Tracing ocean circulation and weathering using radiogenic isotopes

Marcus Gutjahr

Tracing ocean circulation and weathering using radiogenic isotopes

An isotope geochemical approach to reconstruct glacial-interglacial changes in ocean circulation and continental runoff

Südwestdeutscher Verlag für Hochschulschriften

Impressum/Imprint (nur für Deutschland/only for Germany)
Bibliografische Information der Deutschen Nationalbibliothek: Die Deutsche Nationalbibliothek verzeichnet diese Publikation in der Deutschen Nationalbibliografie; detaillierte bibliografische Daten sind im Internet über http://dnb.d-nb.de abrufbar.

Alle in diesem Buch genannten Marken und Produktnamen unterliegen warenzeichen-, marken- oder patentrechtlichem Schutz bzw. sind Warenzeichen oder eingetragene Warenzeichen der jeweiligen Inhaber. Die Wiedergabe von Marken, Produktnamen, Gebrauchsnamen, Handelsnamen, Warenbezeichnungen u.s.w. in diesem Werk berechtigt auch ohne besondere Kennzeichnung nicht zu der Annahme, dass solche Namen im Sinne der Warenzeichen- und Markenschutzgesetzgebung als frei zu betrachten wären und daher von jedermann benutzt werden dürften.

Coverbild: www.ingimage.com

Verlag: Südwestdeutscher Verlag für Hochschulschriften GmbH & Co. KG
Heinrich-Böcking-Str. 6-8, 66121 Saarbrücken, Deutschland
Telefon +49 681 37 20 271-1, Telefax +49 681 37 20 271-0
Email: info@svh-verlag.de

Approved by: Zurich, ETH, Diss. ETH Nr. 16763, 2006

Herstellung in Deutschland (siehe letzte Seite)
ISBN: 978-3-8381-3351-5

Imprint (only for USA, GB)
Bibliographic information published by the Deutsche Nationalbibliothek: The Deutsche Nationalbibliothek lists this publication in the Deutsche Nationalbibliografie; detailed bibliographic data are available in the Internet at http://dnb.d-nb.de.

Any brand names and product names mentioned in this book are subject to trademark, brand or patent protection and are trademarks or registered trademarks of their respective holders. The use of brand names, product names, common names, trade names, product descriptions etc. even without a particular marking in this works is in no way to be construed to mean that such names may be regarded as unrestricted in respect of trademark and brand protection legislation and could thus be used by anyone.

Cover image: www.ingimage.com

Publisher: Südwestdeutscher Verlag für Hochschulschriften GmbH & Co. KG
Heinrich-Böcking-Str. 6-8, 66121 Saarbrücken, Germany
Phone +49 681 37 20 271-1, Fax +49 681 37 20 271-0
Email: info@svh-verlag.de

Printed in the U.S.A.
Printed in the U.K. by (see last page)
ISBN: 978-3-8381-3351-5

Contents

Abstract...4

Zusammenfassung..8

Acknowledgements 13

1. Introduction..15

1.1   Present-day ocean circulation 16

1.2   Classical paleoceanographic proxies 18

1.3   The Last Glacial Cycle in the North Atlantic 19

1.4   Sources and distribution of trace metals in seawater..............................20

1.5   Formation of Fe-Mn oxyhydroxide coatings 23

1.6   Outline of the Thesis..25

2   Reliable extraction of a deepwater trace metal isotope signal from Fe-Mn
     oxyhydroxide coatings of marine sediments...27

2.1   Introduction..28

2.2   Material and Methods 31

2.3   Results...37

    2.3.1   Strontium isotope results...37
    2.3.2   Thorium isotope results...38
    2.3.3   Osmium isotope results..39
    2.3.4   Rare earth elements...39
    2.3.5   Elemental ratios 41

2.4   Discussion...44

    2.4.1   Mass balance calculations...44
    2.4.2   Thorium isotopes in Fe-Mn oxyhydroxide coatings...................49
    2.4.3   Rare earth element patterns...51
    2.4.4   Neodymium and lead mass balance...52

2.5   Conclusions...55

3 Neodymium isotope evolution of North Atlantic Deep and Intermediate Waters in the western North Atlantic since the Last Glacial Maximum 57

3.1   Introduction 59

3.2   Material and Methods 63

    3.2.1   Nd isotopes 63
    3.2.2 $^{230}Th_{xs}$ and sediment redistribution 65

3.3   Results 67

    3.3.1   Flow dynamics, sedimentation rates and focusing 67
    3.3.2   Neodymium isotope results 70

3.4   Discussion 78

    3.4.1   Present day Nd isotope distribution at the Blake Ridge 78
    3.4.2   A plausible mechanism to export a surface water signal 79
    3.4.3   Deep water Nd isotope evolution 80
    3.4.4   GNAIW Nd isotope composition 81
    3.4.5   Reduced export of NADW to the Southern Ocean (Cape Basin) during the LGM? 83

3.5   Conclusions 84

4 Seawater Pb isotopes in the western North Atlantic: Laurentide ice sheet decay, continental freshwater runoff diversions and the establishment of the Holocene weathering regime 86

4.1   Introduction 88

4.2   Material and Methods 90

4.3   Results 95

    4.3.1   Glacial-interglacial $^{208,207,206}Pb/^{204}Pb$ isotope evolution 95
    4.3.2   Combined seawater Pb isotope records 98
    4.3.3 $^{207}Pb/^{206}Pb$ and $^{208}Pb/^{206}Pb$ isotope trends 100
    4.3.4   Combined glacial-interglacial Pb and Nd isotope evolution ... 102

4.4   Discussion 103

    4.4.1   Florida current contributions to the NADW Pb isotope signal ... 104
    4.4.2   A mean $^{207}Pb/^{206}Pb$ age from Fe-Mn oxyhydroxides coatings ... 105
    4.4.3   Reduction of riverine Pb input during LGM? 106
    4.4.4   Agreement between LGM and pre-Pleistocene Pb isotopes? ... 107
    4.4.5   Chemical weathering on the North American continent 108
    4.4.6   Persistent continental runoff diversion into the western North Atlantic 110

4.5   Synthesis and Conclusions 114

5   **A first record of glacial-interglacial Hf isotope variations in seawater at sub-millennial resolution** 117

5.1   Introduction 119

5.2   Material and Methods 122

5.3   Results 125

    5.3.1   Grain-size effects 125
    5.3.2   Elemental ratios 127
    5.3.3   Mass balance calculations 129
    5.3.4   Seawater Hf isotope trends since the LGM 132
    5.3.5   Pb-Nd-Hf isotope trends since the LGM 134
    5.3.6   Pb-Nd-Hf isotope trends in core 31GGC, 3410 m 135
    5.3.7   Neodymium-hafnium trends 137
    5.3.8   North Atlantic water column ε_{Hf} variability during the LGM 139

5.4   Discussion 140

    5.4.1   Applicability of the mass balance calculations 140
    5.4.2   Seawater ε_{Hf} trends since the LGM 142
    5.4.3   The "dissolved Hf" issue 146

5.5   Conclusions 147

6   **General conclusions and outlook** 150

6.1   Conclusions 151

    6.1.1   New proxy development in sedimentary archives 151
    6.1.2   Major North Atlantic water mass changes traced with Nd isotopes .. 153
    6.1.3   Constraining the timing of continental runoff reorganisations from eastern North America 154
    6.1.4   Glacial-interglacial Hf weathering trends 156

6.2   Outlook 157

    6.2.1   Specific suggestions for future research 157
    6.2.2   General suggestions for future research 159

References 161

Appendix 179

Age Calibration 180

Abstract

This dissertation aims to develop, evaluate and use new radiogenic isotopic proxies for paleoceanographic and paleoclimatic reconstructions at sub-millennial resolution. Fe-Mn oxyhydroxide coatings in marine pelagic sediments are an ideal archive for this purpose because seawater-derived trace metals such as Nd, Hf and Pb can be chemically extracted from these coatings. This archive was already successfully used in earlier studies for the reconstruction of the seawater Nd isotope evolution in the South Atlantic, but no such record at sub-millennial resolution exists for the North Atlantic. For Hf and Pb isotopes, not a single record employing Fe-Mn oxyhydroxide coatings exists to date. Especially Hf and Pb isotopes in seawater, however, potentially yield valuable information about short-term climatic changes, intensity of glaciation and the prevailing weathering regime on the continents. Therefore, using seawater-derived Hf and Pb isotopes in conjunction with Nd isotopes from the same seawater-derived fraction allow determining the provenance of a water mass and the climatic conditions prevailing at the continental source area.

The first part of the thesis evaluates the chemical signature of the extracted Nd and Pb isotope signal from Fe-Mn oxyhydroxide coatings. The ^{87}Sr/^{86}Sr compositions, used in earlier studies as evidence for the seawater origin of extracted Nd from Fe-Mn oxyhydroxide fractions, indicated detrital contributions to the seawater signal for many samples studied here. The Pb and Nd isotope compositions on the other hand provided consistent results, for which reason a series of tests were carried out to monitor the reliability of the ^{87}Sr/^{86}Sr isotope signal. The extracted and the residual detrital phase were analysed for their (a) Rare Earth Element (REE) patterns, (b) elemental ratios such as Al/Nd and Al/Pb. Additionally, (c) mass balance calculations were carried out. Those were used to determine the detrital contributions to the respective Sr, Nd and Pb isotope composition. The results provided compelling evidence that the extracted Fe-Mn oxyhydroxide fraction is indeed chemically very distinct from the residual detrital phase. Furthermore, seawater Nd and Pb isotope compositions can be reliably extracted despite ^{87}Sr/^{86}Sr offset from seawater ratios. This is due to the fact that a significant portion of bulk Nd and Pb in marine sediments is of seawater origin, whereas relatively little seawater-derived Sr is incorporated into Fe-Mn oxyhydroxide fractions compared with bulk sediment Sr concentrations.

The second aim of this thesis is to characterize the water column stratification along the Blake Ridge today, during the Last Glacial Maximum (LGM), and to track the hydrographic changes along the shallow, intermediate and deeper Blake Ridge. The typical North Atlantic Deep Water (NADW) Nd isotope composition (ε_{Nd} ~13.5) could be measured for water depths within the major flow path of the Deep Western Boundary Current (DWBC). Above the major flow axis of the DWBC Nd isotope compositions of Fe-Mn oxyhydroxide coatings were biased towards surface water compositions, most likely caused by re-distribution of North American shelf sediment, which carried pre-formed authigenic Fe-Mn oxyhydroxide coatings downslope the Blake Ridge. This sediment re-distribution phenomena is also indicated by highly elevated $^{230}Th_{xs}$, reflecting significant lateral addition of sediment towards the shallow sites of the Blake Ridge during the Holocene. Throughout the LGM and the deglaciation, the characteristic interglacial Lower NADW is not observable along the deeper Blake Ridge at all based on Nd isotopes, and is only initiated after the Younger Dryas. The results of this study indicate that the Glacial North Atlantic Intermediate Water (GNAIW) is isotopically up to 3.5 to 4 ε_{Nd} more radiogenic than the interglacial NADW. This suggests that Labrador Sea Water did not contribute to GNAIW during the LGM.

The Pb isotopic evolution recorded along the Blake Ridge in the transition from the LGM towards the Holocene reveals climatic trends that are closely coupled with the retreat of the Laurentide ice sheet. Controlling factors are climate and the freshwater runoff from the North American continent, whereas source only plays a minor role. Lead is released incongruently during early chemical weathering, which is reflected in very radiogenic Pb isotope compositions during incipient chemical weathering. The initial radiogenic Pb isotopic pulse during incipient chemical weathering likely reflects the interplay of (a) the efficient washout of loosely bound alpha-recoiled radiogenic Pb from damaged lattices sites of freshly crushed rock substrate during the Last Glacial Maximum. Furthermore, (b) continuing chemical weathering can supply very radiogenic Pb through the preferential dissolution of U- and Th-rich accessory mineral phases. In every studied sediment core the $^{206}Pb/^{204}Pb$, $^{207}Pb/^{204}Pb$ and $^{208}Pb/^{204}Pb$ seawater isotope compositions changed from unradiogenic compositions

5

during the LGM and most of the deglaciation to extremely radiogenic compositions at around 11.2 calendar years BP. The rise towards very radiogenic compositions starts approximately with the beginning of the Younger Dryas (YD) at ~13 ka BP. The most radiogenic Pb isotope compositions were recorded at 11.2 ka BP, hence post-dating the YD, rapidly dropping again towards intermediately radiogenic Pb isotope compositions today. The switch from unradiogenic to extremely radiogenic Pb isotope compositions occurred over the course of the YD and continued for several hundred years afterwards. This finding is important as it evidences major continental freshwater drainage changes that persisted until after the end of the YD. Prior to ~13 ka BP, continental runoff was directed into the Gulf of Mexico, for which reason no earlier continental warming signal could be recorded along the Blake Ridge. Significant volumes of continental runoff were directly drained into the western North Atlantic after 13 ka BP.

The Hf isotope signal extracted from Fe-Mn oxyhydroxide fractions along the Blake Ridge offered additional information to weathering-related aspects on the North American continent. In contrast to Nd and Pb, the purity of the seawater Hf isotope signal could not be unambiguously assessed because of its very different elemental behaviour. Hafnium is depleted in Fe-Mn oxyhydroxide coatings relative to Nd and Pb and only small concentrations could be extracted during reductive leaching. A few Fe-Mn oxyhydroxide fractions produced Hf isotope compositions, which seem to represent leaching artefacts. On the other hand, the majority of the extracted Fe-Mn oxyhydroxide coatings produced consistent results that agree with ferromanganese crust compositions from the western North Atlantic, furthermore displaying comparable Al/Hf elemental ratios to ferromanganese crust compositions from the abyssal Pacific. Both the deeper and the shallow Blake Ridge follow a Hf isotopic evolution from unradiogenic ε_{Hf} during the LGM to radiogenic ε_{Hf} today. Lowest ε_{Hf} values were recorded in the deeper Blake Ridge during the LGM, yielding ε_{Hf} as low as -3.1. Hafnium isotope compositions become increasingly more radiogenic immediately following the LGM and the Younger Dryas is marked by a short-lived excursion towards slightly less radiogenic compositions. The Hf isotope trends seen along the Blake Ridge support the zircon-grinding effect proposed in earlier studies and suggest weathering trends from more congruent glacial erosion during the last

6

glacial towards a more zircon-free incongruent bulk weathering Hf input into the North Atlantic during the Holocene. In the light of missing direct seawater Hf isotope compositions, however, this first glacial-interglacial Hf isotope seawater record must remain tentative.

Overall the results presented here highlight the potential of using Fe-Mn oxyhydroxide coatings as paleoceanographic archives in a combined Nd, Hf and Pb isotopic approach. This dissertation only covers Marine Isotope Stages 1 and 2 in the western North Atlantic and certainly much more can be learned from this archive, extending the records back in time and by applying them to different key locations.

Zusammenfassung

Diese Dissertation widmet sich der Entwicklung, Evaluation und der Anwendung neuer radiogener Isotopensysteme für paläozeanographische und paläoklimatische Rekonstruktionen mit einer zeitlichen Auflösung von unter tausend Jahren. Eisen-Mangan-Oxyhydroxide in marinen pelagischen Sedimenten stellen ein ideales Archiv dar für diese Zwecke. Spurenmetalle wie zum Beispiel Nd, Hf oder Pb lagern sich in diesem Meerwasser-Archiv ein und können chemisch extrahiert werden. Obwohl dieses Archiv in früheren Studien schon erfolgreich zur Rekonstruktion der Nd-Isotopie des Tiefenwassers im Südatlantik angewendet wurde, existiert noch keine solche hoch auflösende Meerwasser-Rekonstruktion der Nd-Isotopie für den Nordatlantik. Für die isotopische Hf- und Pb-Entwicklung des Meerwassers gibt es noch keinerlei hochauflösende Datensätze. Jedoch besonders die Hf- und Pb-Isotopie des Meerwassers kann möglicherweise wertvolle Informationen über kurzfristige Klimaschwankungen, den Umfang eiszeitlicher Vergletscherung und des vorherrschenden Verwitterungsregimes auf den Kontinenten enthalten. Deshalb ermöglicht die gekoppelte Anwendung der Hf- und Pb-Isotopien der aus dem Meerwasser abgelagerten Fe-Mn Oxyhydroxide, in Verbindung mit deren Nd-Isotopie die Bestimmung der Herkunft einer Wassermasse, und darüber hinaus eine Abschätzung der klimatischen Bedingungen im kontinentalen Herkunftsgebiet.

Der erste Teil der Arbeit evaluiert den Charakter der Nd- und Pb-Fraktion, die aus Fe-Mn Oxyhydroxiden extrahiert werden. Die ^{87}Sr/^{86}Sr-Verhältnisse derselben Proben, die in früheren Studien als Beleg der Meerwasserherkunft der Spurenmetalle angewandt wurde, deutete für extrahiertes Nd und Pb in dieser Arbeit detritische Verunreinigungen des isotopischen Meerwassersignals an. Die Nd- und Pb-Isotopie andererseits waren sehr konsistent und reproduzierbar, aus welchem Grund eine Reihe an Tests durchgeführt wurden, um die Stichhaltigkeit der ^{87}Sr/^{86}Sr-Verhältnisse zu beurteilen. Die extrahierten und verbleibenden Phasen der marinen Sedimente wurden (a) auf deren Seltene-Erden-Charakeristika, (b) Elementverhältnisse wie zum Beispiel Al/Nd und Al/Pb untersucht. Ausserdem wurde (c) eine Massenbilanzberechnung erstellt, um den Effekt detritischer Beiträge zum Sr-, Nd- und Pb-Meerwasser-Isotopensignal zu quantifizieren. Die Ergebnisse dieser Tests liefern eindrückliche Beweise, dass die extrahierte Meerwasser-Oxyhydroxid-Fraktion sich tatsächlich chemisch eindeutig von der detritischen Fraktion unterscheidet. Ferner kann ein Meerwasser Isotopensignal sehr wohl für Nd und Pb extrahiert werden, obwohl die Sr-Isotopie vermeintlich signifikante detritische Verunreinigungen andeutet. Dies liegt daran, dass sich ein erheblicher Teil des gesamten Nd und Pb in marinen Sedimenten aus dem Meerwasser darin ablagerte, während – verglichen mit detritischen Sr-Konzentrationen – nur relativ wenig Sr aus dem Meerwasser in jenen Meerwasserablagerungen eingebaut wird.

Das zweite Ziel dieser Arbeit ist die Charakterisierung der Wassermassen-Stratifizierung entlang des Blake Ridge im westlichen Nordatlantik heute und jener des letzten Glazials (LGM). Darüber hinaus sollen Zirkulations-Änderungen zeitlich aufgelöst werden, die sich seit der letzten Eiszeit entlang des Blake Ridge ereigneten. Eine für Nordatlantisches Tiefenwasser (NADW) typische Nd-Isotopie (ε_{Nd} ~13.5) konnte in Wassertiefen entlang der Hauptfliessachse dieses Tiefenwassers in 3200 m Wassertiefe und darunter gemessen werden. In Wassertiefen oberhalb der Hauptfliessachse des NADW schien die extrahierte Nd-Isotopie durch Beiträge aus der oberen Wassersäule von der tatsächlichen Tiefenwasser-Isotopie abzuweichen. Der Grund dafür liegt wahrscheinlich an Sediment-Umverteilungsprozessen. Aufgewirbeltes Sediment vom amerikanischen Schelf exportiert so vorgeformte Fe-Mn Oxyhydroxide während dieser Umverteilung in tiefere Abschnitte des Blake Ridge. Dieses Sediment-Umverteilungs-Phänomen wird durch erheblich erhöhte $^{230}Th_{xs}$-Werte während des Holozäns belegt. Während des letzten Glazials und bis in die Jüngere Dryas hinein ist die charakteristische NADW-Tiefenwasserzirkulation entlang des tieferen Blake Ridge nicht nachweisbar. Laut der Nd-Isotopie setzt die moderne Tiefenwasserzirkulation erst nach der Jüngeren Dryas ein. Die hier vorgelegten Ergebnisse deuten an, dass das glaziale nordatlantische Zwischenwasser (GNAIW) sich in dessen Nd-Isotopie deutlich vom interglazialen NADW unterscheidet und etwa 3.5 bis 4 ε_{Nd} radiogener ist. Aufgrund dieser Beobachtung leistete in der Labrador-See gebildetes Tiefenwasser vermutlich keinen Beitrag zur Bildung des GNAIW während des letzten glazialen Maximums.

Die isotopische Meerwasser-Entwicklung gelösten Bleis entlang des Blake Ridges im Übergang vom Letzten Glazialen Maximum zum Holozän zeigt klimatische Trends auf, die eng mit dem Rückzug des Laurentidischen Eisschildes in Nordamerika im Zusammenhang stehen. Kontrollierende Parameter sind vor allem das vorherrschende Klima und der (Schmelz-) Wasserabfluss vom nordamerikanischen Kontinent. Dagegen spielt die kontinentale Herkunft des gelösten Bleis nur eine untergeordnete Rolle. Blei wird in frühen Stadien der chemischen Verwitterung inkongruent gelöst, was zur Folge hat, dass während der frühen chemischen Verwitterung eine extrem radiogene Pb-Isotopie gelöst wird. Der initiale radiogene Pb-Schub während der frühen chemischen Verwitterung ist wahrscheinlich hauptsächlich (a) einerseits auf das effiziente Lösen von locker gebundenem radiogenen Pb zurückzuführen, das durch Alpha-Zerfall erzeugt wurde und so aus dem geordnetem Kristallgitter der Minerale herauskatapultiert wurde. Fortschreitende chemische Verwitterung hat aber auch (b) das bevorzugte Lösen verwitterungsunbeständiger U- und Th-reicher akzessorischer Minerale zur Folge, die wiederum hochradiogenes Blei in den Verwitterungskreislauf abgeben. Das frische Gesteins-Substrat hierfür wurde in der vorhergehenden Kaltzeit mechanisch zerkleinert. In jedem analysierten Sedimentkern entlang des Blake Ridge änderte sich $^{206}Pb/^{204}Pb$, $^{207}Pb/^{204}Pb$ and $^{208}Pb/^{204}Pb$ von unradiogenen Verhältnissen während des letzten Glazials und weiten Teilen des Deglazials hin zu extrem radiogenen Verhältnissen nach der Jüngeren Dryas vor etwa 11'200 Kalenderjahren. Die Änderung von unradiogenen hin zu extrem radiogenen Pb-Isotopenverhältnissen begann zeitgleich mit der Jüngeren Dryas vor etwa 13'000 Jahren. Die radiogenste Pb-Isotopie trat vor etwa 11'200 Jahren auf und fällt somit nicht mehr in die Jüngere Dryas. Die Meerwasser-Pb-Isotopie fiel danach schnell wieder hin zu intermediär radiogenen Verhältnissen heute. Der Wechsel von unradiogenen zu extrem radiogenen Pb-Isotopenverhältnissen entwickelte sich stetig über die Jüngere Dryas hinweg und stieg auch nach dem Ende der Jüngeren Dryas noch für einige hundert Jahre weiter an. Dieses Feststellung ist wichtig da sie unmittelbar die Neugliederung der Wasserabfluss-Pfade vom Amerikanischen Kontinent widerspiegelt. Vor dem Beginn der Jüngeren Dryas floss kontinentales (Schmelz-) Frischwasser über den Mississippi in den Golf von Mexiko, weshalb zu jener Zeit noch keine Änderung der Meerwasser-Pb-Isotopie am Blake Ridge festgestellt werden konnte. Erst als der kontinentale Frischwasserabfluss zu Beginn der Jüngeren Dryas direkt in den Nordatlantik umgelenkt wurde, konnte dies im Meerwasser Pb-Isotopensignal am Blake Ridge aufgelöst werden.

Die aus Eisen-Mangen-Oxyhydroxiden aus Sedimenten des Blake Ridge gewonnene Hf-Isotopie birgt zusätzliche Informationen zu Verwitterungsprozessen in Nordamerika. Im Gegensatz zu Nd und Pb konnte die Reinheit des Hf-Meerwasser-Isotopensignals aufgrund dessen sehr unterschiedlichen chemischen Verhaltens nicht eindeutig belegt werden. Hafnium ist abgereichert in Fe-Mn Oxyhydroxiden verglichen mit Nd oder Pb und nur geringe Hf-Konzentrationen konnten aus dieser Phase gelöst werden während des reduzierenden chemischen Lösens. Manche Fe-Mn Oxyhydroxide ergaben Ausreisser, die vermutlich Anlösungs-Artefakte der detritischen Fraktion widerspiegeln. Andererseits ergab der Grossteil der gelösten Fe-Mn Oxyhydroxide sehr konsistente und reproduzierbare Hf-Isotopien, die mit hydrogenetischen Eisenmangankrusten-Zusammensetzungen im Nordatlantik übereinstimmen. Weiterhin konnten Al/Hf-Verhältnisse in jenen Fe-Mn Oxyhydroxiden gemessen werden, die mit hydrogenetischen Eisenmangankrusten aus dem zentralen Südpazifik übereinstimmen. Meerwasser entlang des tiefen, wie auch des flacheren Blake Ridge zeigen einen glazial-interglazialen Trend von unradiogener Hf-Isotopie während des letzten glazialen Maximums hin zu radiogener Hf-Isotopie heute. Niedrigste ε_{Hf} mit Werten von -3.1 wurden entlang des tieferen Blake Ridge im letzten Glazial gemessen. Die Hf-Isotopie des Meerwassers wurde zunehmend radiogener nach dem letzten glazialen Maximum, und sogar die Jüngere Dryas ist durch eine kurzzeitig auftretende unradiogene Variation aufgelöst. Die zeitlichen Trends in der Hf-Isotopie des Meerwassers liefern weitere Indizien für den in früheren Studien an Eisenmangankrusten vorgeschlagenen Zirkon-Verwitterungseffekt. Der am Blake Ridge beobachtetete Trend zeugt vermutlich von kongruenterer glazialer Erosion während des letzten Glazials hin zu inkongruenter „Zirkon-freier" Gesamtgesteins-Verwitterung während des Holozäns. Da noch keine direkte Meerwasser Hf-Isotopendaten publiziert wurden, bedarf dieser erste glazial-interglaziale Meerwasser-Datensatz noch Bestätigung durch direkte Meerwasser-Hf-Isotopendaten.

Die in dieser Arbeit vorgestellten Ergebnisse belegen das immense Potential der Nutzung der Nd-, Hf- und Pb-Isotopie von Fe-Mn Oxyhydroxiden, die aus marinen Sedimenten gelöst werden können für paläozeanographische und paläoklimatische Studien. Diese Dissertation konzentriert sich ausschliesslich auf die Marinen

Isotopenstadien 1 und 2 im westlichen Nordatlantik und mit Sicherheit birgt dieses Archiv noch viel mehr Information. Dafür sollten die Datensätze weiter in die Vergangenheit erweitert, aber auch an andern Schlüssel-Lokationen der Weltmeere angewendet werden.

Acknowledgements

First of all I want to say thank you to Martin for being a great supervisor, not only for the scientific guidance but also for support in those little minor political issues I had the joy to experience through picking the working site along the Blake Ridge. Whenever there were manuscripts or other things to correct or to comment on he was always quick in returning great in-depth reviewed drafts. Learning from Martin was very inspiring, both in terms of efficiency and scientific approach. Going out for a beer after work with him was obviously also a great thing to do!

Alex has been a great co-supervisor, and especially in the past few months I very much enjoyed getting feedback from him. Being part of Alex's work group to carry out a PhD project was a great scientific and social experience. The Isotope Geology group at ETH was huge when I arrived here and I learned about isotope systems and scientific applications, which I would have not imagined before ETH.

Sidney, it was a great pleasure to have you as an external examiner! Whenever there were conferences at which I presented my latest results I always enjoyed the scientific discussions arising when you came to see the posters. Judy, I hope you still can enjoy PhD exams having to read about two theses per month at the moment. Thank you very much for becoming my internal official supervisor after the departure of Martin and Alex.

Our lab and the machines simply would not work if we had not such a great team of technicians. Even if the mass specs broke down Friday after lunch they would be back working before the evening. Many thanks to Urs Menet, Andreas Süsli, Donat Niederer, Heiri Baur and Felix Oberli for keeping the machines running smoothly, as well as for supplying specially made little gadgets that we needed for doing lab work. Mark Rehkämper, Helen Williams, Claudine Stirling, Sarah Woodland, Ben Reynolds, Sonja Ripperger and Thorsten Kleine did a great job in maintaining the mass specs. Heiri Baur and Bruno Rütsche maintain an outstanding computer support. Thank you Marie-Theres for all the help in the lab. Britt and Valentina were always a great help when anything administrative had to be done. Morten, Emma-Kate and Martin Wipf spent quite some time introducing me to Th isotope extraction and measurements. Thanks also to Tina van de Flierdt for taking over some of the lab introductory work at the beginning of my thesis - and for many good scientific discussions. Several written communications with Lloyd Keigwin also improved the quality of this thesis. He highlighted some very interesting aspects from a different paleoceanographic perspective.

Although Bernard was not involved with my PhD research he was very supportive in letting me presenting latest results at EGU this year. Thank you for being the chair in my defense.

The best about ETH are the dozens of PhD students running around here! Thanks to all the people that made the past four years a great lively experience, including all the PhD students and postdocs from our group and the rest of the department. Special thanks to all the "mates" during those years at ETH, including Adélie Delacour, Agnès Markowski, Alex Teague, Andrew Stewart, Ansgar Grimberg, Bastian Georg, Ben Reynolds, Caroline Harris, Chris Krugh, Christian Liebske, Darrell Harrison, Elena Melekhova, Emmanuel Chapron, Erwan Le Guerroué, Helen Williams, DJ Jogi Rickli (the sound machine), Léo Luzieux, Louise Gall, Luca Carricchi, Maarten Aerts, Manuela Fehr, Mathieu Touboul, Morten Andersen, Paola Ardia, Pauline Rais, Philipp Heck, Pierre Bouilhol, Raquel Alonso, Sarah[3] (Woodland-Aciego-Bureau), Sebastien Castelltort, Stefan Strasky, Sune Nielsen, Sonja Ripperger, Tina van de Flierdt, Tonny Thomsen, Veronika Klemm, Zarina Minubayeva and Zheng Zhou and Samuele Toschini. The generations of Friday beer organizers deserve ETH medals. Ansgar was always very helpful when events had to be organized. Stephan Teuwissen from the Theaterwerkstatt Dynamo in Zürich helped me keeping a non-scientific artistic eye open.

Chapter 1

Introduction

1.1   Present-day ocean circulation

In simplified terms today's interhemispheric thermohaline circulation pattern can be described as being driven by sinking of cold dense salty water at high latitudes in the North Atlantic exporting freshly formed North Atlantic Deep Water (NADW) into the Southern Ocean where it joins the Antarctic Circumpolar Current (ACC) (Fig. 1.1) (Broecker, 1991; Schmitz, 1995). The feature of water subduction in the North Atlantic is caused by a yearly net vapour loss of about 0.32 Sv (1 Sv = 10^6 m^3s^{-1}) in the North Atlantic (Broecker, 1991), which is compensated by salt export. This mode of meridional overturning circulation pattern has been stable for the past 10 kyr, but probably was fundamentally different during the last glacial cycle (Broecker et al., 1985; Sarnthein et al., 1994; Ganopolski and Rahmstorf, 2001) and possibly even switched off temporarily during catastrophic iceberg discharge Heinrich event 1 and the Younger Dryas (Broecker et al., 1988b; McManus et al., 2004; Rickaby and Elderfield, 2005; Robinson et al., 2005).

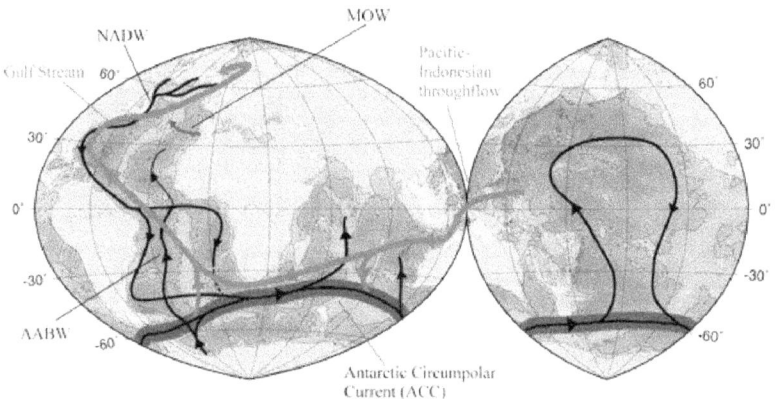

Figure 1.1. Simplified illustration of the present-day thermohaline circulation, driven by sinking of saline cold water in the North Atlantic. Red arrows indicate surface currents, black arrows deep water circulation. NADW: North Atlantic Deep Water; MOW: Mediterranean Outflow Water; AABW: Antarctic Bottom Water (from Frank, 2002).

The freshwater flux from the continents into the North Atlantic is a crucial parameter in this regard because excessive freshwater fluxes could dilute Gulf Stream water (representing proto-NADW) in so far that sinking of salty water stops and the ocean conveyor halts. Indeed it has been argued that the short-lived Younger Dryas cold interval (12.9 to 11.5 ka BP) (Hughen et al., 2004b) was potentially triggered by the Lake Agassiz meltwater diversion from the Mississippi drainage system into the Gulf of St. Lawrence. Thereby North Atlantic water would have been diluted to the extent that it could not sink anymore during cooling after its transit from low to high latitudes within the Gulf Stream (Broecker et al., 1988b; Broecker et al., 1989a). Evidence for increased meltwater fluxes in the Younger Dryas were also found in sediment cores adjacent to SE Greenland (Jennings et al., 2006).

Modern NADW can be identified throughout the Atlantic by various parameters. It is nutrient-depleted (high $\delta^{13}C$, low phosphate (PO_4) concentrations) (Fig. 1.2) (Broecker, 1991; Charles and Fairbanks, 1992) and yields young radiocarbon ventilation ages because of its recent contact and exchange with the atmosphere (Broecker et al., 1988a). For comparison, water in the deep northern Pacific, representing the end of the deep ocean conveyor, is nutrient-enriched and displays oldest radiocarbon ventilation ages.

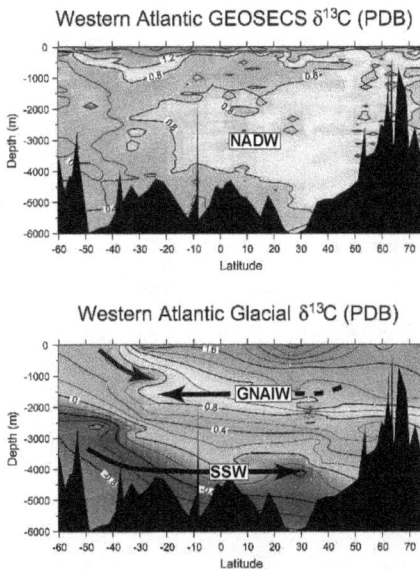

Figure 1.2. Atlantic water mass stratification today and during the LGM as inferred from stable carbon isotope records. NADW: North Atlantic Deep Water; GNAIW: Glacial North Atlantic Intermediate Water; SSW: Southern Source Water (modified from Curry and Oppo, 2005).

During the Last Glacial Maximum (LGM) the interglacial NADW was replaced by Glacial North Atlantic Intermediate Water (GNAIW) (Fig. 1.2), whereas the entire deeper Atlantic was probably under the influence of Southern Source Water (SSW) advecting from the Southern Ocean (Labeyrie et al., 1992; Curry and Oppo, 2005). The exact spatial extent of SSW advection into the North Atlantic, as well as the change from a glacial to the modern circulation mode is unresolved at present. Similarly unclear are the causes that triggered the major hydrographical changes in the North Atlantic basin. Stable carbon isotopes, the classical paleoceanographic proxy used to decipher water mass properties, failed to faithfully record these changes in the North Atlantic (e.g., Keigwin, 2004).

1.2   Classical paleoceanographic proxies

The vast majority of knowledge about past climates and paleo-circulation patterns in the oceans is derived from stable oxygen and carbon isotopic analyses. The $^{18}O/^{16}O$ composition of calcite shells in marine organisms such as benthic or planktonic foraminifera provide information about ambient water temperature and the isotopic composition of seawater, and also contains information about ice volume (Shackleton et al., 1984; Chappell and Shackleton, 1986; Shackleton, 1987). Oxygen isotope compositions are generally given relative to a standard (i.e., either VPDB for marine organisms or V-SMOW for ice core records, cf. Fig. 1.3) in per mil notation ($\delta^{18}O$).

The stable carbon isotopic compositions of benthic foraminifera reflect changes in the ratio of dissolved inorganic carbon of ambient seawater (e.g., Zachos et al., 2001). Nutrient depletion of seawater due to high productivity results in high $^{13}C/^{12}C$ ratios in marine organisms such as foraminifera. Organisms preferentially incorporate ^{12}C, leading to relative enrichment of the residual water with ^{13}C. Because NADW represents former Gulf Stream water it can be identified because of its high $\delta^{13}C$ compositions (Fig. 1.2) (e.g., Duplessy et al., 1988; Charles and Fairbanks, 1992; Vautravers et al., 2004).

Besides stable carbon isotopes, Cd/Ca have been employed for paleoceanographic reconstructions because the elemental ratios trace nutrient contents of seawater similarly to $\delta^{13}C$ (Boyle, 1988; Boyle, 1992). While stable $\delta^{13}C$ isotope or Cd/Ca variations induced by biological processes helped to identify water mass properties in the past no quantitative estimates for water mass mixing can be made using these classical proxies because they are affected by productivity and water temperature (e.g. Marchitto and Broecker, 2006). Here radiogenic isotope compositions of trace metals such as Nd, Hf and Pb allow better insights because these trace metals are unaffected by biologically induced fractionation and more reliably reflect the provenance and flow path of a water mass (Frank, 2002).

1.3   The Last Glacial Cycle in the North Atlantic

Figure 1.3 shows Northern Greenland (NGRIP) ice core stable oxygen isotope record spanning the last glacial cycle (Andersen et al., 2004a). Compared with the fully glacial Marine Isotope Stages 4 and 2 (MIS 4: ~78-64 ka BP; MIS 2: ~32-17 ka BP), the oxygen isotope record during the present MIS 1 suggest significantly warmer (interglacial) and stable climatic conditions. The short-term climatic excursion during the Younger Dryas is also resolved in the Northern Greenland ice core (Fig. 1.3).

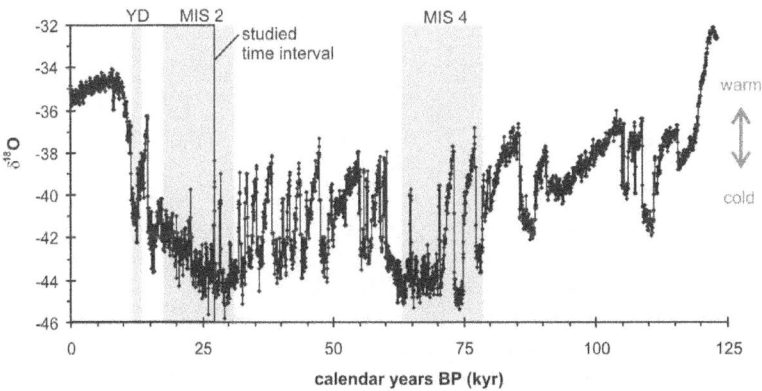

Figure 1.3. Northern Greenland (NGRIP) ice core stable oxygen isotope record spanning the last glacial cycle. Isotopic ratios (δ^{18}O) are expressed in per mil with respect to Vienna Standard Mean Ocean Water (V-SMOW). The time interval studied in this thesis is indicated by the black rectangle. (created with data from Andersen et al., 2004a).

The extreme climatic conditions during the LGM resulted in extensive growth of the Laurentide and Innuitian ice sheets (Fig. 1.4) (Dyke et al., 2002), which only retreated slowly during the deglaciation (Dyke and Prest, 1987; Licciardi et al., 1998). Although the Labrador Sea was ice covered for a significant part of the year, it was seemingly not shielded with a perennial sea ice cover (de Vernal and Hillaire-Marcel, 2000).

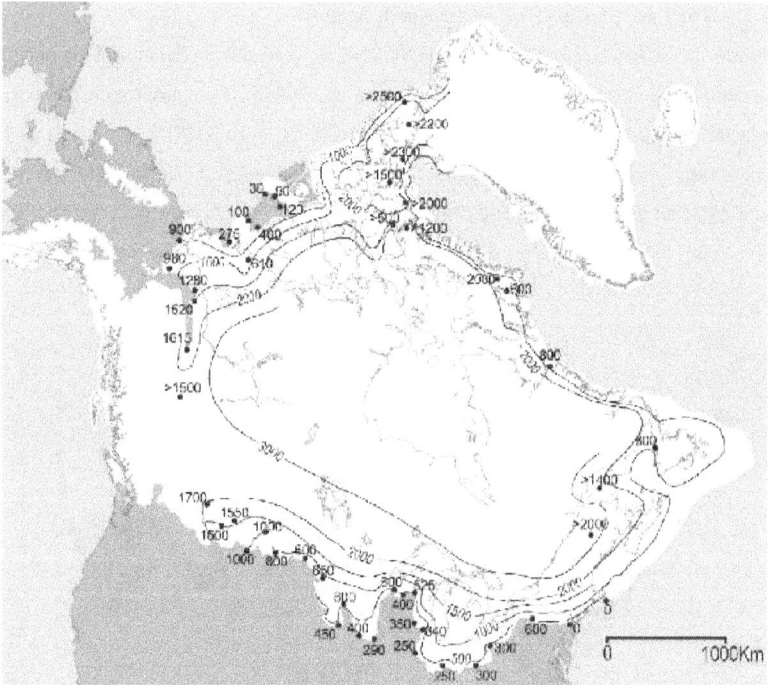

Figure 1.4. Extent of the Laurentide and the Innuitian ice sheets during the LGM as reconstructed by Dyke et al. (2002). Ice surface contours are based on elevations along the LGM ice margin and topographic high points overridden by ice.

1.4   Sources and distribution of trace metals in seawater

Because the properties of the trace metals Nd, Hf and Pb are introduced in the individual chapters, only a brief overview is given here. These elements have in common that their residence time in seawater is on the order of or shorter than the rate of meridional overturning circulation, which is about 1,500 years (Broecker and Peng, 1982). The variability in their isotopic composition is dominantly a function of the age of the continental source area because all isotope ratios of interest vary with the age of the crustal source area due to radioactive decay. Elemental differentiation between parent and daughter nuclide during continental crust formation on the one hand, and the time elapsed since crystallisation on the other therefore governs the bulk Nd, Hf and Pb isotopic composition of crustal domains (Table 1.1). The Pb and Hf isotope signals released during weathering are in part also strongly climate-dependent.

Element	Parent Isotope	Daughter Isotope	Half-Life	Ratio of Interest
Nd	^{147}Sm	^{143}Nd	106 Gyr	^{143}Nd/^{144}Nd
Hf	^{176}Lu	^{176}Hf	36 Gyr	^{176}Hf/^{177}Hf
Pb	^{238}U	^{206}Pb	4.47 Gyr	^{206}Pb/^{204}Pb
	^{235}U	^{207}Pb	707 Myr	^{207}Pb/^{204}Pb
	^{232}Th	^{208}Pb	14 Gyr	^{208}Pb/^{204}Pb
				^{208}Pb/^{206}Pb
				^{207}Pb/^{206}Pb

Table 1.1. Summary of the long-lived isotopic tracers used in this dissertation. Although for Nd and Hf only the displayed isotopic ratios are variable on Earth, for Pb also the ^{208}Pb/^{206}Pb and ^{207}Pb/^{206}Pb yield information about the provenance of water masses. The reference isotopes ^{144}Nd, ^{177}Hf and ^{204}Pb are primordial isotopes.

Lead displays the shortest residence time (50-400 years) (Craig et al., 1973; Schaule and Patterson, 1981; Cochran et al., 1990), whereas Nd and Hf behave quasi-conservative in seawater with possibly similar residence times ranging from 600 to 2000 years (Jeandel, 1993; Jeandel et al., 1995; Godfrey et al., 1997; Lee et al., 1999; Tachikawa et al., 1999; David et al., 2001). Neodymium and Hf isotopes are therefore suitable tracers for water exchange between the major ocean basins, whereas Pb isotope compositions are the preferred proxy to detect local input sources (Figs. 1.5, 1.6, 1.7). The main input sources for Nd, Hf and Pb are the continents (Frank, 2002), although a hydrothermal contribution has been invoked to partially control the Hf budget in seawater (White et al., 1986; Bau and Koschinsky, 2006). Hydrothermal Pb input sources can be found locally close to mid-ocean ridges (van de Flierdt et al., 2004c).

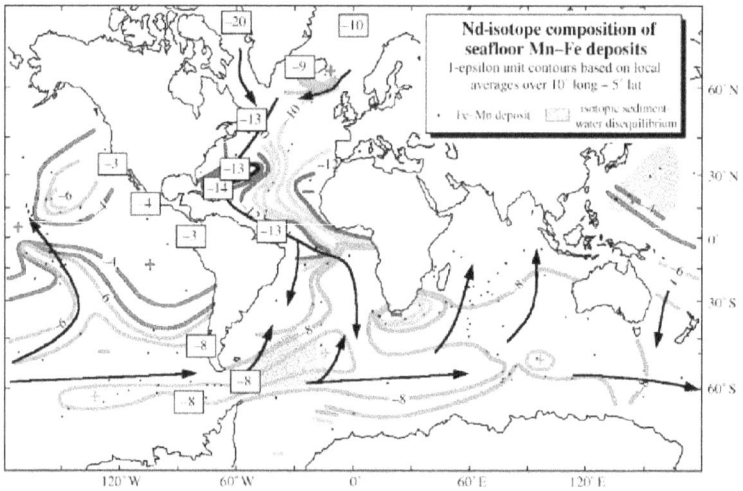

Figure 1.5. Map of Nd isotope variability in ferromanganese deposits (given ε_{Nd} notation; $\varepsilon_{Nd} = (^{143}Nd/^{144}Nd_{sample}/^{143}Nd/^{144}Nd_{CHUR} - 1) * 10^4)$) as shown by Goldstein and Hemming (2003). Least radiogenic Nd isotopic compositions are found in hydrogenetic deposits in the western North Atlantic, most radiogenic compositions in the North Pacific.

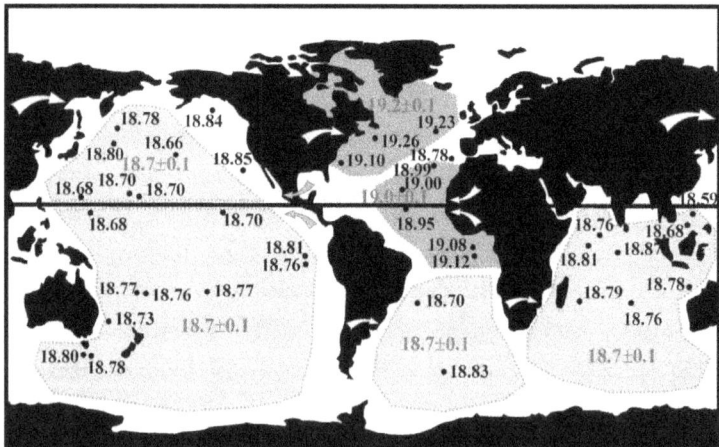

Figure 1.6. Present-day $^{206}Pb/^{204}Pb$ isotopic composition of ferromanganese deposits (Klemm et al., 2007). The most radiogenic Pb isotopic compositions are found in hydrogenetic deposits in the western North Atlantic, least radiogenic compositions in the equatorial and North Pacific. Compared with crustal Pb isotopic compositions the different abyssal ocean basins display remarkably homogeneous compositions.

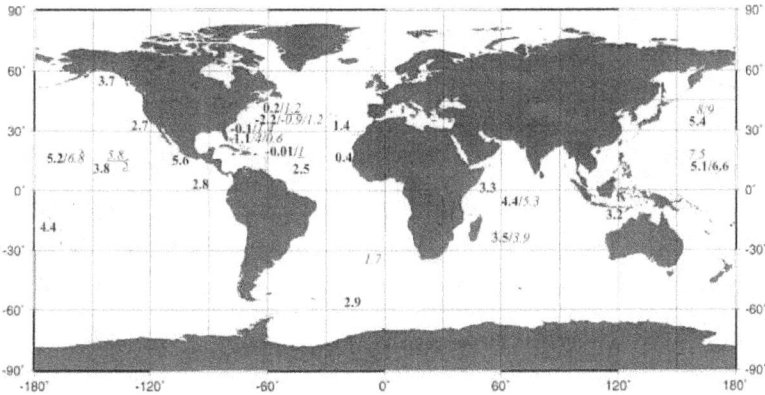

Figure 1.7. Present-day Hf isotope compositions (given ε_{Hf} notation; ε_{Hf} = $(^{176}Hf/^{177}Hf_{sample}/^{176}Hf/^{177}Hf_{CHUR} - 1) * 10^4$) of surface scrapings in ferromanganese crusts and nodules compiled by David et al. (2001).

In the western North Atlantic this is particularly useful because the NADW, which occupies most of the western North Atlantic below circa 1000 m water depth entrains Labrador Sea water, which drains the old cratonic Canadian shield and therefore yields very distinct Nd, Hf and Pb isotopic compositions. Because Nd (Hf) is more incompatible than Sm (Lu), it was more efficiently incorporated into the Canadian shield during formation, for which reason this old cratonic province has very low bulk $^{143}Nd/^{144}Nd$ and $^{176}Hf/^{177}Hf$ (Figs. 1.5 and 1.7). For comparison, young circum-Pacific island arc rocks have completely different isotopic compositions (Figs. 1.5 and 1.7).

Whereas Nd is released more or less congruently from continental crust, with the implication that the dissolved riverine load will reflect the bulk continental source Nd isotope composition, this is not the case for Hf and Pb. These elements are released incongruently to different extent during chemical weathering. The mechanism behind this incongruent release will be introduced and discussed in chapters 4 (for Pb) and 5 (for Hf).

1.5 Formation of Fe-Mn oxyhydroxide coatings

Amorphous Fe-Mn oxyhydroxides, the paleoceanographic archive employed in this dissertation, develop in the pore waters of marine sediments (Fig. 1.8). In order to identify seawater isotope fluctuations at highest possible resolution high sedimentation rates are needed and intervals with low sedimentation rates (< 5 cm/kyr) are expected to release a somewhat time-integrated seawater isotope signal.

Rare Earth Element (REE) pore water profiles suggest that under oxic conditions trace metals are scavenged from pore waters and incorporated into Fe-Mn oxyhydroxide coatings within the uppermost few centimetres below seafloor (Haley et al., 2004). Compared with the overlying water mass Haley et al. (2004) found highest REE concentrations in the upper 5 cm of marine sediments studied, which indicates that most trace metals are possibly incorporated into Fe-Mn oxyhydroxides within this zone. Haley et al. (2004) also illustrated the effect of pore waters anoxia, when Fe-Mn oxyhydroxides are dissolved leading to trace metal concentrations in the pore waters, which are an order of magnitude higher than those under oxic conditions (see also Elderfield and Sholkovitz, 1987; Thomson et al., 1993b; Thomson et al., 1995). Therefore, in studies employing the Fe-Mn oxyhydroxide fraction as paleoceanographic archives it is important to assess whether pore water anoxia prevailed at any time obliterating the original seawater signal.

It was also suggested that the original authigenic Fe-Mn oxyhydroxide signal in marine sediments can be disturbed through pre-formed Fe-Mn oxides in detrital material (Bayon et al., 2004). These authors concluded that the authigenic Fe-Mn oxyhydroxide signal in a sediment core from the Angola basin was biased through pre-formed detrital Fe oxide component imparted to the sediment site through Congo River material. Conversely, Bayon et al. (2004) also found that Fe-Mn oxyhydroxides extracted from sediment cores further south in the Cape Basin are of authigenic origin. These findings suggest that the reliability of the Fe-Mn oxyhydroxide radiogenic isotopic signal has to be evaluated on a site specific basis.

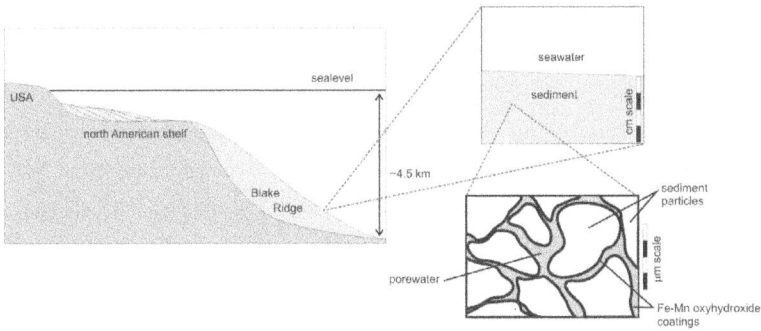

Figure 1.8. Schematic illustration showing the formation of dispersed Fe-Mn oxyhydroxide coatings in the pore waters of marine sediments, thereby incorporating seawater-derived trace metals that carry the radiogenic (Nd, Hf, Pb, Th) isotopic composition of the overlying water mass.

1.6   Outline of the Thesis

This dissertation comprises four individual papers, which present and discuss geochemical and isotopic aspects concerning the extraction of the authigenic Fe-Mn oxyhydroxide fraction from marine sediments, give an insight into hydrographic re-organisations in the North Atlantic since the LGM, and further trace paleoclimatic and weathering-related trends on the North American continent.

Chapter 2 serves to corroborate the applicability to extract a pure seawater signal from marine sediments in order to determine its radiogenic (Nd, Pb, Th) isotope composition back in time. The extraction method is introduced and evaluated in detail and a strong case is made for the reliability of the extracted signal.

Neodymium isotope compositions extracted from Fe-Mn oxyhydroxide fractions are reliable recorders to track the provenance of water masses and are employed in chapter 3 to identify the timing of the establishment of the present-day circulation mode in the western North Atlantic. Because the Blake Ridge is a drift deposit the issue of re-distributed sediment needs to be addressed, and $^{230}Th_{xs}$ results are presented that illustrate such sediment re-distribution phenomena and their implications on the ambient seawater signal incorporated into the Fe-Mn oxyhydroxide coatings.

25

In chapter 4 high-resolution seawater Pb isotope evolution patterns from the same locality since the LGM are presented that resolve a yet undetected short-term radiogenic Pb excursion beginning at the onset of the Younger Dryas, and which traces major continental runoff reorganisations from the North American continent. Moreover, the Pb isotope records trace climatic and weathering trends in North America. This record is the first of its kind and illustrates the potential of extracting a seawater Pb isotope signal from marine sediments. Due to its short residence time in seawater the most pronounced Pb isotopic variations in seawater could not be resolved in earlier studies using ferromanganese crusts because of their slow growth rate.

Chapter 5 addresses the feasibility to extract a seawater Hf isotope signal from Fe-Mn oxyhydroxide fractions. This part of the dissertation was technically the most delicate because of very low Hf concentrations in the extracted seawater fraction and its susceptibility to contamination by partial leaching of the detrital phase. Although a clear case could be made for the seawater origin of Nd, Pb and Th isotopes in chapter 2, the definite proof cannot be served for Hf isotopes. Some leaching artefacts were produced but overall the Hf isotopic trends observed on glacial-interglacial timescales are systematic and agree with ferromanganese crust data, containing valuable information about (sub-) glacial erosion during the LGM.

Chapter 2[*]

Reliable extraction of a deepwater trace metal isotope signal from Fe-Mn oxyhydroxide coatings of marine sediments

[*]submitted to *Chemical Geology* as: Gutjahr, M. H., Frank, M., Stirling, C. H., Klemm, V., van de Flierdt, T. and Halliday, A. N. Reliable extraction of a deepwater trace metal isotope signal from Fe-Mn oxyhydroxide coatings of marine sediments.

Abstract

The extraction of a deepwater radiogenic isotope signal from marine sediments is a powerful though under-exploited tool for the characterisation of past climates and modes of ocean circulation. The radiogenic and radioactive isotope compositions (Nd, Pb, Th) of ambient deepwater are stored in early diagenetic Fe-Mn oxyhydroxide coatings in marine sediments, but the unambiguous separation of the isotopic signal in this phase from other sedimentary components is difficult and measures are needed to ensure its seawater origin. Here the extracted Fe-Mn oxyhydroxide phase is investigated geochemically and isotopically in order to constrain the potential and the limitations of the reconstruction of deepwater radiogenic isotope compositions from marine sediments.

Our results show that the isotope compositions of elements such as Sr and Os obtained from the Fe-Mn oxyhydroxide fraction are easily disturbed by detrital contributions originating from the extraction process, whereas the seawater isotope compositions of Nd, Pb and Th can be reliably extracted from marine sediments in the North Atlantic. The main reason is that the Nd, Pb and Th concentrations in the detrital phase of pelagic sediments are much lower than in the Fe-Mn oxyhydroxide fractions, which is reflected by very low Al/Nd, Al/Pb and Al/Th ratios of the Fe-Mn oxyhydroxide fractions. Mass balance calculations illustrate that the use of the $^{87}Sr/^{86}Sr$ isotope composition to confirm the seawater origin of the extracted Nd, Pb and Th isotope signals is misleading. Even though the $^{87}Sr/^{86}Sr$ in the Fe-Mn oxyhydroxide fractions is often higher than the seawater Sr isotope composition, the corresponding detrital contribution does not translate into altered seawater Nd, Pb and Th isotope compositions due to mass balance constraints.

2.1   Introduction

The chemical extraction (leaching) of the seawater-derived Nd, Pb and Th isotope signal in the early diagenetic Fe-Mn oxyhydroxide phase of marine sediments offers great potential for paleoceanographic and paleoclimatic purposes because it can yield essential information for sub-millennial fluctuations of marine environmental conditions. The benefit of using these seawater-derived radiogenic or radioactive trace metal isotope signatures is that they are not biased by biological processes, unlike other stable isotope or elemental ratio proxies. The analytical approach, however, is not trivial and an inherent caveat of the leaching of Fe-Mn oxyhydroxide coatings from marine sediments is the lack of unambiguous evidence for its unbiased deepwater origin.

Due to its residence time in seawater of 600 to 2,000 years, Nd isotopes are increasingly used as a quasi-conservative water mass tracer for the present and past ocean (Frank, 2002; Goldstein and Hemming, 2003; Piotrowski et al., 2005). Lead is removed much faster from the water column with an average residence time of only about 50 years in the Atlantic (Henderson and Maier-Reimer, 2002). Therefore, dissolved Pb isotope records are suitable archives for the detection of changes in local input sources (Frank, 2002; van de Flierdt et al., 2003). Dissolved Th isotopes in seawater are, as yet, a largely unemployed tool for the determination of changes of particle fluxes from the continents to the oceans because the $^{232}Th/^{230}Th$ is strongly controlled by the presence or absence of particulates in seawater (Guo et al., 1995; Robinson et al., 2004). Considering the repeated deposition of ice rafted debris (IRD) layers during Heinrich events over the last glacial cycle, $^{232}Th/^{230}Th$ might serve as a good proxy for the correlation of North Atlantic sediment cores in the absence of conspicuous IRD layers.

Rare Earth Element (REE) porewater profiles of marine pelagic sediments suggest that under oxic to suboxic conditions trace metals (Sr, Nd, Os, Pb, Th) are scavenged from porewaters and incorporated into Fe-Mn oxyhydroxide coatings within the uppermost few centimetres below seafloor (Haley et al., 2004). These authors found most elevated REE concentrations in the upper 5 cm below seafloor, which indicates that most trace metals are incorporated into Fe-Mn oxyhydroxides within this zone or slightly below. Haley et al. (2004) also illustrated the effect of anoxic porewaters, where Fe-Mn oxyhydroxides are dissolved leading to trace metal concentrations in the pore waters, which are an order of magnitude higher than those found under oxic conditions (see also Elderfield and Sholkovitz, 1987; Thomson et al., 1993b; Thomson et al., 1995). Therefore, in studies employing the Fe-Mn oxyhydroxide fraction as paleoceanographic archives it is important to assess whether pore water anoxia prevailed at any time, which may have obliterated the original seawater signal.

It was also suggested that locally the original authigenic Fe-Mn oxyhydroxide signal in marine sediments can be disturbed through pre-formed Fe-Mn oxides on detrital material (Bayon et al., 2004). These authors concluded that the authigenic Fe-Mn oxyhydroxide signal in a sediment core from the Angola basin was biased through pre-formed detrital Fe oxide component associated with Congo river material transferred to the sedimentation site (Bayon et al., 2004).

Measures are required to assess the reliability of the chemically extracted authigenic trace metal isotope signal. For this purpose, in previous studies using the Nd isotope composition of Fe-Mn oxyhydroxide coatings in pelagic sediments the Sr isotope composition of the same fraction was measured to confirm the pure seawater origin of the Nd in the oxyhydroxide leachate (Rutberg et al., 2000; Piotrowski et al., 2004; Piotrowski et al., 2005). Strontium is a conservative element in seawater with a residence time on the order of two million years (Henderson et al., 1994). The Sr isotope composition extracted from Fe-Mn oxyhydroxides in pelagic sediments of the past 100 kyr should thus reproduce the present-day seawater $^{87}Sr/^{86}Sr$ of 0.70918 if no significant Sr was leached from the detrital fraction. This test was successfully used in earlier studies carried out in the Southern Atlantic, where a seawater Sr isotope composition was indeed measured in the leachates. In contrast, the application of this monitor indicated significant detrital contributions to the leachates of Fe-Mn oxyhydroxides from several sites in the North Atlantic (Piotrowski, 2004). These offsets in Sr isotopes were not accompanied by obvious alterations of the Nd isotope compositions. Using core-top Nd isotope compositions measured in Fe-Mn oxyhydroxide leaches in conjunction with the dissolved SiO_2 concentration of the overlying water mass Piotrowski (2004) observed very good agreement with direct seawater measurements with only two exceptions located near the African coast and on the Mid-Atlantic Ridge flank proximal to the Azores islands. Hence the good agreement in Nd isotope compositions between direct seawater measurements and leached core-top sediments presented by Piotrowski (2004) from the North Atlantic despite non-hydrogenetic Sr isotope compositions measured in the same leachates indicates that the Sr isotope composition as a sole guideline for the assessment of the seawater origin of the Nd isotope composition is probably too strict.

In this study we demonstrate why, regardless of the Sr or Os isotopic compositions, a reliable Nd isotopic composition of seawater can be extracted from the Fe-Mn oxyhydroxide coatings of marine sediments in the North Atlantic, thus allowing the reconstruction of past water mass signatures. Secondly, it is shown that Pb, as well as Th isotope compositions in the same Fe-Mn oxyhydroxide fractions also reflect ambient seawater compositions. The complex paleoceanographic and paleoclimatic implications of the seawater Nd and Pb isotope evolution trends obtained from sediments along the Blake Ridge in the western North Atlantic will be discussed in chapters 3 and 4.

2.2   Material and Methods

Sediment cores along the Blake Ridge in the western North Atlantic have been recovered during R/V Knorr cruise 140 (KNR140). Samples used in this study were taken from core 51GGC (1790 m water depth; 32°N, 76°W) and 12JPC (4250 m water depth; 29°N, 73°W). In order to avoid blank contributions induced trough sieving and successive rinsing of the sediments only bulk sediments were processed. Sediments were dried and ground for homogenisation. The method applied for the extraction of the seawater-derived Fe-Mn oxyhydroxide fraction was modified and combined from a selection of published leaching procedures (Chester and Hughes, 1967; Tessier et al., 1979; Ruttenberg, 1992; Rutberg et al., 2000; Bayon et al., 2002; Tovar-Sanchez et al., 2003; Piotrowski et al., 2004) and is illustrated in Figure 2.1. A commented compilation of various alternative chemical extraction methods can be found in Bayon et al. (2002).

Samples were processed in pre-cleaned 50 ml centrifuge tubes. Carbonate was removed using a Na acetate buffer in a shaker capped with Parafilm® to allow degassing of carbon dioxide during carbonate dissolution. After centrifugation and decanting of the supernatant, loosely adsorbed metals were removed using a 1M $MgCl_2$ solution. Following centrifugation and triple rinses in deionised water (Milli-Q system), the Fe-Mn oxyhydroxide coatings were dissolved by leaching the samples for three hours in a shaker at room temperature in a 0.05M hydroxylamine hydrochloride (HH) – 15 % distilled acetic acid – 0.03M Na-EDTA solution, buffered to pH 4 with analytical grade NaOH.

31

```
┌─────────────────────────────┐
│   ground bulk sediment      │
│        (300 mg)             │
└─────────────────────────────┘
┌─────────────────────────────┐
│    remove carbonate         │
│  1M Na acetate mixed 52:48  │
│ with 1M acetic acid in shaker│
└─────────────────────────────┘
┌─────────────────────────────┐
│ centrifuge & decant supernatant │
└─────────────────────────────┘
┌─────────────────────────────┐
│  leach exchangeable fraction │
│       10 ml 1M MgCl₂        │
└─────────────────────────────┘
┌─────────────────────────────┐
│ centrifuge & decant supernatant │
└─────────────────────────────┘
┌─────────────────────────────┐
│        triple rinse         │
│   MQ grade deionised water  │
│(fill container - agitate in vortex mixer)│
└─────────────────────────────┘
┌─────────────────────────────┐
│ centrifuge & decant supernatant │
└─────────────────────────────┘
┌─────────────────────────────────┐
│  leach Fe-Mn oxyhydroxide fraction │
│ 15 ml 0.05M hydroxylamine hydrochloride- │
│          15 % acetic acid-      │
│           0.03M Na-EDTA         │
│     buffered to pH 4 with NaOH  │
│ at room temperature in shaker for 2-3 hours │
└─────────────────────────────────┘
              ↓
  centrifuge & transfer supernatant
```

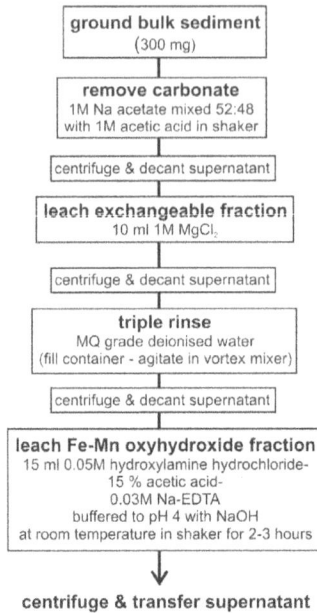

Figure 2.1. Analytical procedure used for the extraction of the Fe-Mn oxyhydroxide coatings.

In addition, the detrital fraction was extracted from the sediments to obtain the complimentary detrital isotope signal for each of the Fe-Mn oxyhydroxide fractions. Because the leaching protocol described above does not extract the Fe-Mn oxyhydroxide fraction quantitatively, a second leach with the above HH-acetic acid-Na-EDTA leach was applied for 24 hours to guarantee complete removal of residual Fe-Mn oxyhydroxide coatings. A further two rinses with deionised water were applied prior to treatment of the samples with *aqua regia* to destroy organic matter. The remaining detrital sediment fraction was dissolved by pressure digestion in a concentrated HF-HNO$_3$ mixture.

The first Fe-Mn oxyhydroxide leach and the detrital fraction were split into separate aliquots for extraction of Sr, Nd and Pb. For a subset of nine Fe-Mn oxyhydroxide coating samples additional aliquots were taken for Th isotope measurements of the Fe-Mn oxyhydroxide fraction, as well as a subset of seven samples for Os isotope analyses.

Separation and purification of the individual elements followed standard procedures, for Sr (Horwitz et al., 1992), Nd (Cohen et al., 1988), Os (Birck et al., 1997; Klemm et al., 2005), Pb (Lugmair and Galer, 1992) and Th (Luo et al., 1997). Total procedural blanks for Sr in the Fe-Mn oxyhydroxide fraction were high due to the use of analytical grade NaOH for buffering of the leach solution, averaging to 22.2 ng and contributing between 0.4 and 8.7 % of the Fe-Mn oxyhydroxide coating signal. The Sr isotope ratios in the mass balance calculation were corrected for this blank contribution, which was found to have no significant effect on our conclusions. Corrected ^{87}Sr/^{86}Sr isotope ratios changed on the third or fourth decimal place (at most 0.0005 for sample 12JPC-55cm), whereas small contributions of the detrital fraction to the leached fraction had much more severe effects on the Sr isotope composition (as much as 0.0128 for 12JPC-20cm, Table 2.2). The total Sr procedural blank for the detrital fraction was 1.7 ng and is less than 0.004 % of the detrital Sr concentration. Procedural blanks for Nd were <30 pg for the oxyhydroxide fraction and <315 pg for the detrital fraction with all blanks being smaller than 0.1% of the total amount of Nd present in each sample. The total procedural blank for Os was below 0.7 pg and the applied blank correction was between 5 and 20 %. For Pb the procedural blank in the Fe-Mn oxyhydroxide fraction was 1.03 ng and always below 0.35 % of the total Pb concentration. In the detrital Pb fraction, the procedural blank was 1.31 ng and below 0.33 % of the total amount of Pb present in the samples. Total procedural blanks for ^{232}Th were <10 pg and were negligible. Procedural blanks for ^{230}Th were below the detection limit of the secondary electron multipliers (SEMs) on the Nu Plasma MC-ICPMS.

Measurements of the Sr, Nd, Pb and Th isotopes were carried out on a Nu Plasma MC-ICPMS at ETH Zürich using the exponential law to correct for instrumental mass fractionation (Rehkamper et al., 2001). Interfering ^{86}Kr (from the Ar carrier gas) and possible ^{87}Rb contributions on the respective Sr masses have been monitored through the simultaneous measurement of ^{83}Kr and ^{85}Rb. Interference-corrected ^{87}Sr/^{86}Sr was adjusted for mass bias by normalising to ^{86}Sr/^{88}Sr of 0.1194 (Nier, 1938; Steiger and Jager, 1977). The elevated scattered background in the Sr mass range was accounted for by subtracting linearly interpolated zero beam intensities on 1/3 masses above and below the mass of interest. All reported Sr isotope results were normalised to ^{87}Sr/^{86}Sr

= 0.710245 for the NIST SRM987 Sr standard. The long-term reproducibility for repeated measurements of the NIST SRM987 Sr standard was 0.000025 (2σ; n=70).

Measured $^{143}Nd/^{144}Nd$ was corrected for the instrumental mass bias applying a $^{146}Nd/^{144}Nd$ of 0.7219. All reported Nd isotope results were normalised to a $^{143}Nd/^{144}Nd$ of 0.512115 for the JNdi-1 standard (Tanaka et al., 2000). The long-term reproducibility for repeated measurements of the JNdi-1 standard was ±0.27 ϵ_{Nd} (2σ; n=70). The $^{143}Nd/^{144}Nd$ results presented here are expressed in standard epsilon notation relative to a Chondrite Uniform Reservoir (CHUR):

$$\varepsilon_{Nd} = \left[\frac{^{143}Nd/^{144}Nd_{sample}}{^{143}Nd/^{144}Nd_{CHUR}} - 1 \right] \times 10^4$$

($^{143}Nd/^{144}Nd_{CHUR}$ = 0.512638 (Jacobsen and Wasserburg, 1980))

Lead isotopes were measured using a Tl-doping procedure (Walder and Furuta, 1993; Belshaw et al., 1998) with a Pb/Tl ~ 4. The expected offset to TIMS Pb isotope data (cf., Thirlwall, 2002) are accounted for by normalising all respective Pb isotope compositions to the triple spike TIMS Pb ratios for NIST SRM 981 reported by Galer and Abouchami (1998).

Thorium isotope compositions ($^{232}Th/^{230}Th$) in the Fe-Mn oxyhydroxide fractions were measured using an external normalization technique by doping with the CRM 145 U metal standard solution (formerly NIST SRM 960). The instrumental mass bias was corrected for by normalising to a $^{238}U/^{235}U$ value of 137.88 (Steiger and Jager, 1977). Individual measurements were split into two cycles due to the cup configuration of the MC-ICPMS (Table 2.1). Using this setup, the minor ^{230}Th ion beam was monitored on SEM 2 (IC2), simultaneously with ^{235}U and ^{238}U on Faraday detectors in the first cycle, while ^{238}U, ^{235}U and ^{232}Th were monitored subsequently on Faraday collectors during a second analysis sequence (Table 2.1). The relative gain between the Faraday collectors and each SEM was calibrated against the CRM 145 U standard solution doped with a $^{233}U/^{229}Th$ spike. All isotope ratios were corrected for the minor amounts of natural ^{238}U, ^{235}U, ^{234}U, ^{232}Th and ^{230}Th present in the artificial spike. After a preceding peak tailing characterisation of the major ^{232}Th ion beam monitoring the mass range between 230.6 and 228.5 on the SEM, the linearly interpolated half mass zeros on ^{230}Th have been corrected by -8.84 %. The Th104 standard supplied by C. Innocent reproduced at 27,260 ± 89 (n=41, 2σ) and the Th105 standard at 217,080 ± 1428 (n=22, 2σ). Abundance sensitivity is on the order of 2.5 to 5 ppm amu^{-1}(cf., Andersen et al., 2004b).

	F(+2)	F(+1)	F(Ax)	F(-1)	F(-2)	IC0(-3)	F(-4)	IC1(-5)	F(-6)	IC2(-7)
Zero 1	238.5	237.5		235.5	234.5	233.5	232.5			229.5
Zero 2	239.5	238.5		236.5	235.5	234.5	233.5			230.5
Cycle 1		^{238}U			^{235}U	^{234}U				^{230}Th
Cycle 2	^{238}U			^{235}U			^{232}Th			

Table 2.1. Collector array of the Nu Plasma MC-ICPMS (Nu1) used for the Th isotope analysis at the ETH, Zürich. Due to the cup configuration the measurements have been split in two cycles, allowing the determination of the ^{230}Th abundance in cycle one and the ^{232}Th abundance in cycle two. Numbers in brackets refer to mass differences (amu) to the axial Faraday collector.

Osmium isotope compositions and the total amount of Os within each sample as determined by isotope dilution, were obtained by negative thermal ionisation mass spectrometry (N-TIMS) following published methods (Creaser et al., 1991; Volkening et al., 1991; Birck et al., 1997). The external reproducibility of the Os isotope measurements over the course of 3 years, determined through repeated measurements of an internal Os-standard was ^{187}Os/^{188}Os = 0.1081 ± 0.0006 (2σ) (n = 18).

On a set of nine sediment samples a small fraction of the first (3h) and the second (24h) Fe-Mn oxyhydroxide leachate, as well as an aliquot of the detrital fraction was separated for the determination of major and trace element concentrations, measured by ICP-OES and ICP-MS at the Geological Institute of the University of Kiel (Germany). The individual Sr, Nd, Os, Pb and Th isotope results are displayed in Table 2.2. Results of the rare earth element analyses are shown in Table 2.3.

Calendar ages used in Fig. 2.2a are based on conventional ^{14}C ages published by Keigwin (2004); transformed into calendar years using the marine radiocarbon age calibration Marine04 of Hughen et al. (2004) assuming ΔR = 0.

Core	Depth in core (cm)		$^{87}Sr/^{86}Sr$ ± 0.000025	$^{143}Nd/^{144}Nd$ ± 0.000014	Nd ± 0.27	$^{187}Os/^{188}Os$ ± 0.0006	Os [pg/g]	$^{206}Pb/^{204}Pb$ ± 0.0016	$^{207}Pb/^{204}Pb$ ± 0.0017	$^{208}Pb/^{204}Pb$ ± 0.0052	$^{208}Pb/^{206}Pb$ ± 0.00015	$^{207}Pb/^{206}Pb$ ± 0.00003	$^{232}Th/^{230}Th$
51GGC	60	coating	0.709181	0.512127	-10.0	1.053 ± 0.005	9.53	19.1337	15.6906	39.1620	2.04678	0.82007	n.d.
51GGC	270	coating	0.709576	0.512114	-10.2	n.d.	n.d.	19.2187	15.6844	39.2735	2.04346	0.81610	n.d.
51GGC	316	coating	0.709488	0.512109	-10.3	1.269 ± 0.014	2.55	19.1229	15.6781	39.1306	2.04630	0.81987	n.d.
51GGC	350	coating	0.709773	0.512088	-10.7	1.154 ± 0.009	7.02	19.0723	15.6672	39.0757	2.04881	0.82147	n.d.
51GGC	390	coating	0.708940	0.512164	-9.3	n.d	n.d.	19.0280	15.6730	38.9298	2.04593	0.82367	n.d.
51GGC	400	coating	0.709835	0.512154	-9.4	1.177 ± 0.006	2.09	18.9804	15.6622	38.8977	2.04942	0.82520	n.d.
12JPC	55	coating	0.715004	0.511963	-13.2	1.081 ± 0.015	1.22	19.3772	15.6928	39.5909	2.04318	0.80986	n.d.
12JPC	85	coating	0.713909	0.512069	-11.1	1.157 ± 0.008	1.69	19.1117	15.6650	39.1725	2.04968	0.81966	n.d.
12JPC	263	coating	0.712391	0.512119	-10.1	1.095 ± 0.014	1.68	18.9438	15.6543	38.9209	2.05454	0.82636	n.d.
51GGC	60	detritus	0.720764	0.512076	-11.0	n.d.	n.d.	18.9916	15.6499	38.8863	2.04755	0.82404	n.d.
51GGC	270	detritus	0.722284	0.512083	-10.8	n.d.	n.d.	19.0187	15.6492	38.9023	2.04548	0.82283	n.d.
51GGC	316	detritus	0.721076	0.512057	-11.3	n.d.	n.d.	18.8819	15.6247	38.7390	2.05165	0.82750	n.d.
51GGC	350	detritus	0.726593	0.512003	-12.4	n.d.	n.d.	18.9437	15.6252	38.8736	2.05206	0.82482	n.d.
51GGC	390	detritus	0.717890	0.512071	-11.1	n.d.	n.d.	18.9218	15.6196	38.7637	2.04863	0.82548	n.d.
51GGC	400	detritus	0.723073	0.512033	-11.8	n.d.	n.d.	19.1015	15.6363	38.9689	2.04010	0.81859	n.d.
12JPC	55	detritus	0.728611	0.511868	-15.0	n.d.	n.d.	18.3156	15.5268	38.5224	2.10325	0.84774	n.d.
12JPC	85	detritus	0.730571	0.511943	-13.6	n.d.	n.d.	18.6927	15.5893	38.8042	2.07590	0.83398	n.d.
12JPC	263	detritus	0.732071	0.512012	-12.2	n.d.	n.d.	18.8683	15.6246	38.9345	2.06349	0.82809	n.d.
51GGC	40	coating	0.711523	0.512117	-10.2	n.d.	n.d.	19.0866	15.6862	39.1302	2.05015	0.82184	43460 ± 70
51GGC	160	coating	0.710463	0.512140	-9.7	n.d.	n.d.	19.1674	15.6912	39.1748	2.04387	0.81865	40100 ± 80
51GGC	300	coating	0.712866	0.512107	-10.4	n.d.	n.d.	19.2333	15.6864	39.2660	2.04152	0.81557	47500 ± 140
51GGC	330	coating	0.711554	0.512091	-10.7	n.d.	n.d.	19.0970	15.6686	39.1061	2.04784	0.82048	58890 ± 220
51GGC	370	coating	0.714968	0.512106	-10.4	n.d.	n.d.	19.0683	15.6648	39.0736	2.04917	0.82152	70090 ± 230
51GGC	371	coating	0.713560	0.512085	-10.8	n.d.	n.d.	19.0678	15.6659	39.0737	2.04919	0.82158	75530 ± 210
51GGC	405	coating	0.716739	0.512132	-9.9	n.d.	n.d.	18.9927	15.6664	38.9289	2.04968	0.82487	63490 ± 180
12JPC	20	coating	0.721992	0.511992	-12.6	n.d.	n.d.	19.2170	15.6811	39.4023	2.05038	0.81600	24130 ± 40
12JPC	233	coating	0.721761	0.512126	-10.0	n.d.	n.d.	18.9306	15.6485	38.8740	2.05349	0.82662	73240 ± 160

Table 2.2. Summary of all isotope results obtained in the course of this study. Sr isotope results for the upper nine Fe-Mn oxyhydroxide coating samples have been corrected for a blank contribution, as well as the Os samples. The lowermost nine Sr isotope results, however, are not corrected for the blank contribution. Os isotopes have been measured with N-TIMS, the remaining isotope systems by MC-ICPMS on a Nu Plasma instrument. Also shown are Os concentrations for the Fe-Mn oxyhydroxide fractions analysed isotopically in this study.

Rare earth element concentrations, given in ng/g of leached bulk sediment.

	La	Ce	Pr	Nd	Sm	Eu	Gd	Tb	Dy	Ho	Er	Tm	Yb	Lu
First Fe-Mn oxyhydroxide leach fraction (3 hours)														
51GGC -														
60 cm	3594	5904	945	3515	736	165	714	111	646	126	318	47	294	42
270 cm	4412	9013	1289	4946	1066	236	1019	157	891	171	423	62	382	54
316 cm	2335	4069	660	2495	530	114	508	75	413	78	190	27	167	23
350 cm	3414	7540	1010	3895	876	187	837	124	676	125	302	44	273	38
390 cm	2014	3572	582	2312	520	117	531	77	435	84	206	29	177	26
400 cm	1984	4246	660	2642	658	146	654	94	506	92	218	30	186	26
12JPC -														
55 cm	3446	15486	1118	4189	927	187	844	126	669	120	285	41	247	34
85 cm	3405	11577	1182	4684	1126	233	1040	151	780	136	311	43	260	36
263 cm	2710	8234	1042	4248	1060	224	954	136	679	114	254	35	209	28
Second Fe-Mn oxyhydroxide leach fraction (24 hours)														
51GGC -														
60 cm	1424	2979	413	1561	328	71	297	44	237	42	96	13	74	10
270 cm	2500	6161	764	2898	588	123	514	76	411	76	183	26	157	21
316 cm	720	1587	213	805	165	33	147	21	111	20	45	6	35	5
350 cm	1756	4303	542	2118	476	95	429	62	320	57	134	19	115	16
390 cm	544	1097	157	611	139	29	130	18	90	16	37	5	31	4
400 cm	657	1497	198	753	165	31	145	19	90	16	36	5	30	4
12JPC -														
55 cm	1765	6201	506	1826	362	67	306	44	228	40	95	13	80	11
85 cm	2223	6509	652	2415	499	94	426	61	303	52	119	17	98	13
263 cm	2339	6524	720	2705	568	108	478	67	326	54	122	17	98	13
Detrital fraction														
51GGC -														
60 cm	24399	42820	5112	17300	2814	510	2244	302	1857	346	970	129	1043	132
270 cm	36607	65166	7871	26630	4414	822	3595	510	3153	608	1699	243	1861	251
316 cm	7562	14664	1710	5973	1044	202	875	128	816	159	453	66	513	70
350 cm	32739	63228	8065	27354	4614	875	3781	551	3394	667	1848	271	2007	276
390 cm	17698	35497	4296	15338	2760	478	2319	325	2091	395	1143	154	1301	168
400 cm	30282	58156	6904	24104	4115	740	3376	481	3126	610	1759	248	2008	265
12JPC -														
55 cm	35010	63651	7325	24830	4009	781	3183	430	2826	529	1542	194	1699	201
85 cm	55414	102186	12050	40821	6558	1258	5203	742	4570	888	2483	358	2704	361
263 cm	37681	81959	11161	39466	7365	1434	5980	906	5624	1116	3080	462	3377	471

Table 2.3. Individual Rare Earth Element concentrations measured in the different phases shown in Figure 2.4. Concentrations are given in ng per gram of bulk sediment used. Note that the second leach, applied for 24 hours in a shaker, was not able to extract more material than the first leach, which was only applied for 3 hours.

2.3   Results

2.3.1   Strontium isotope results

Many of the leached Fe-Mn oxyhydroxide fractions shown in Table 2.2 yielded $^{87}Sr/^{86}Sr$ that is offset from the expected seawater value of 0.70918 (Henderson et al., 1994). Generally $^{87}Sr/^{86}Sr$ offsets were higher in the deeper core than in the shallow core, and the offset from seawater $^{87}Sr/^{86}Sr$ is roughly inversely correlated with grain size of the bulk sediment (not shown). The most radiogenic Sr was found in the deep core ($^{87}Sr/^{86}Sr = 0.72199$ for JPC12 – 20 cm). Hence for the vast majority of the Fe-Mn oxyhydroxide leachates the $^{87}Sr/^{86}Sr$ suggests a significant detrital Sr contribution to the leached fraction (Table 2.2). The Fe-Mn oxyhydroxide fractions contained between 870 to 2450 ng Sr per gram of leached sediment (Table 2.5). Only coating sample 51GGC - 60 cm contained relatively high Sr concentrations in the Fe-Mn oxyhydroxide fraction (14.06 μg/g of leached Sr in Fe-Mn oxyhydroxide fraction). Compared with the sum of total Sr present in both the Fe-Mn oxyhydroxide- and detrital fractions, only about 0.5 to 3.9 % of the total Sr in the sediments is seawater-derived (with the exception of 51GGC - 60 cm, in which 19.4 % of total Sr in the oxyhydroxide fraction originated from seawater).

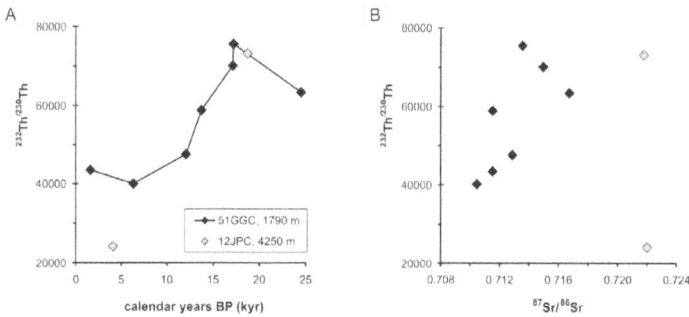

Figure 2.2.   (A) Th isotope time series for core 51GGC in 1790 m, and two additional results from core 12 JPC in 4250 m. The cores show the same decrease in Th isotope composition during the transition from the Deglaciation to the Holocene. (B) Th and Sr isotope cross-plot that reveals no systematic coupling between these isotope systems. If the Th isotope composition was controlled by leaching of the detrital fraction, a pronounced co-variation should be observable. 2σ error bars are smaller than symbol size.

2.3.2   Thorium isotope results

In the shallow core 51GGC the decay-corrected $^{232}Th/^{230}Th$ in the leached Fe-Mn oxyhydroxide fractions ranges from 43,460 in Holocene samples to 75,530 during the earliest deglaciation (17.2 ka BP; Fig. 2.2 and Table 2.2). Deep core 12JPC shows a similar pattern for two measured samples. The Holocene sample at 20 cm depth yielded a decay-corrected $^{232}Th/^{230}Th$ of 24,130 and one sample representing the latest Last Glacial Maximum gave 73,240. The $^{232}Th/^{230}Th$ ratios presented here are in the range of present-day $^{232}Th/^{230}Th$ values published from nearby locations close to Cape Hatteras north of Blake Ridge, which range from $^{232}Th/^{230}Th$ of 122,840 in surface waters to $^{232}Th/^{230}Th$ of 27,990 in 750 m water depth (Guo et al., 1995). Seawater $^{232}Th/^{230}Th$ compositions for the Last Glacial Maximum and the deglaciation are not available for comparison.

2.3.3   Osmium isotope results

Only one out of seven Fe-Mn oxyhydroxide fractions yielded the present-day seawater $^{187}Os/^{188}Os$ of ~1.06 (Levasseur et al., 1998; Peucker-Ehrenbrink and Ravizza, 2000). In Figure 2.3 the $^{187}Os/^{188}Os$ of core 51GGC shows a shift towards more radiogenic continental $^{187}Os/^{188}Os$ accompanied by only minor offsets from seawater $^{87}Sr/^{86}Sr$. Results from deep core 12JPC indicate less detrital contribution to the authigenic $^{187}Os/^{188}Os$, which, however, correspond to significant offsets in $^{87}Sr/^{86}Sr$. The large offset from seawater $^{187}Os/^{188}Os$ is explainable by the low Os concentrations in the respective aliquots, ranging from 1.22 to 9.53 pg per gram of leached bulk sediment (Table 2.2). The very low abundance of Os in the oxyhydroxide fraction reflects its very low seawater concentrations and is responsible for the high susceptibility of the leached isotope ratios to detrital contamination.

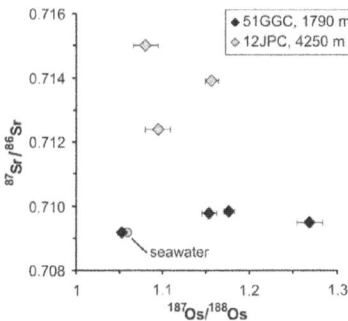

Figure 2.3.   Sr and Os isotope cross-plot, demonstrating the offset for the majority of samples from seawater compositions. Only one sample corresponds to present day seawater.

2.3.4   Rare earth elements

Figure 2.4 shows the results for rare earth element (REE) analyses carried out on leachates of a set of 9 samples from the two core locations. The PAAS-normalised REE plot of the two Fe-Mn oxyhydroxide fractions show a distinct mid REE (MREE) enrichment (Fig. 2.4a and 2.4b), which is not evident in the detrital fraction (Fig. 2.4c). This MREE enrichment is essentially identical to marine pore water REE patterns where particulate Fe^{3+} oxides are reduced and dissolved, thereby releasing scavenged REE (Haley et al., 2004). In contrast and according to expectations, the detrital phases of the same sediment samples display flat PAAS-normalised REE patterns (Fig. 2.4c).

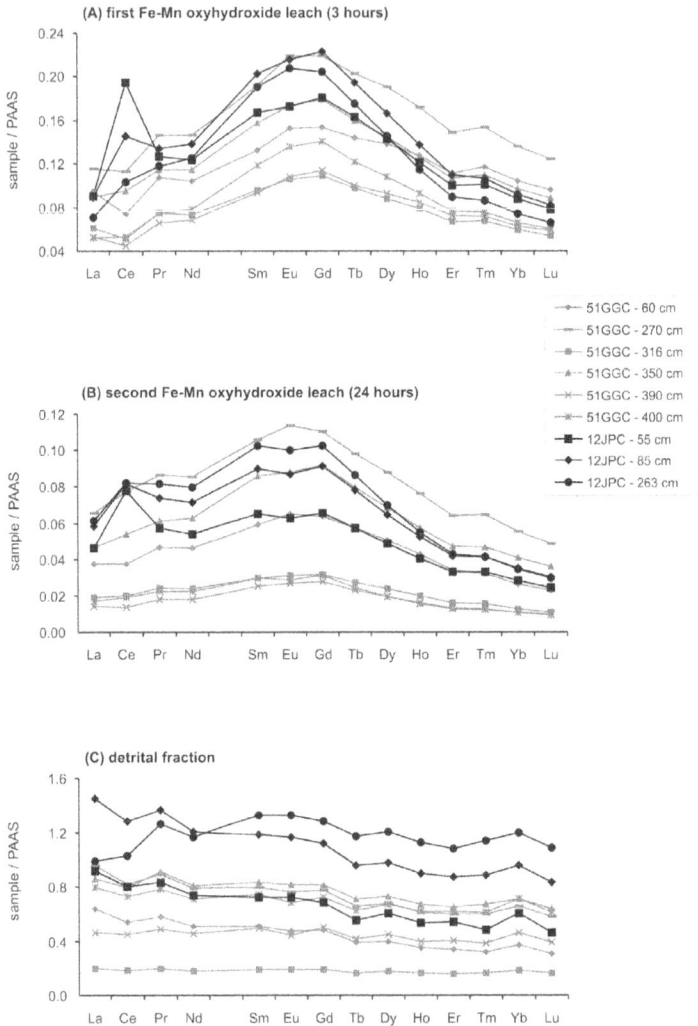

Figure 2.4. Post-Archean Australian average shale normalised (PAAS) REE multi-element plots for two sequential Fe-Mn oxyhydroxide leaching phases and the respective detrital fraction. The second leach was applied to ensure complete removal of the Fe-Mn oxyhydroxide phase. PAAS data are from McLennan (1989).

Al, Nd, Pb and Th concentrations of ferromanganese crusts and sediment fractions

	Al(mg/g)[1,2]	Nd (ppm)	Pb (ppm)	Th (ppm)	Al/Nd	Al/Pb	Al/Th
Central Equatorial Pacific [1]							
Marshall Is.	11.9	170	1799	10.2	70	7	1167
NW of Marshall Is.	12.0	238	1689	-	50	7	-
Johnston I.	14.2	210	1871	17.7	68	8	802
South Pacific [1]	15.8	226	741	2.1	70	21	7535
Atlantic [1]	31.6	-	1133	54.2	-	28	583
				Average:	**65**	**11**	**3168**
THIS STUDY	Al(μg/g) [3]	Nd (μg/g) [3]	Pb (μg/g) [3]	Th (μg/g) [3]	Al/Nd	Al/Pb	Al/Th
First Fe-Mn oxyhydroxide leach (3 hours)							
51GGC - 60 cm	36.1	3.52	1.37	1.11	10.3	26.4	32.7
51GGC - 270 cm	49.0	4.95	2.18	1.94	9.9	22.5	25.3
51GGC - 316 cm	24.7	2.50	1.14	1.04	9.9	21.7	23.7
51GGC - 350 cm	50.4	3.89	2.63	1.48	12.9	19.2	34.0
51GGC - 390 cm	11.9	2.31	1.10	0.86	5.2	10.9	13.9
51GGC - 400 cm	31.0	2.64	1.79	1.45	11.7	17.4	21.5
12JPC - 55 cm	97.7	4.19	6.35	2.66	23.3	15.4	36.8
12JPC - 85 cm	82.2	4.68	6.12	2.55	17.6	13.4	32.2
12JPC - 263 cm	77.0	4.25	4.32	2.58	18.1	17.8	29.9
				Average:	**13.2**	**18.3**	**28**
Second Fe-Mn oxyhydroxide leach (24 hours)							
51GGC - 60 cm	85	1.56	0.74	0.63	55	116	137
51GGC - 270 cm	103	2.90	1.53	1.10	36	67	93
51GGC - 316 cm	48	0.81	0.97	0.33	59	49	145
51GGC - 350 cm	87	2.12	2.16	0.93	41	40	93
51GGC - 390 cm	53	0.61	0.39	0.23	87	134	226
51GGC - 400 cm	80	0.75	0.43	0.29	107	185	281
12JPC - 55 cm	116	1.83	2.09	1.24	64	56	94
12JPC - 85 cm	122	2.41	2.78	1.50	51	44	81
12JPC - 263 cm	146	2.70	2.21	1.66	54	66	88
				Average:	**61**	**84**	**138**
Detrital fraction							
51GGC - 60 cm	7640	17.30	6.41	6.17	442	1192	1238
51GGC - 270 cm	11054	26.63	11.89	10.70	415	930	1033
51GGC - 316 cm	6408	5.97	2.70	1.87	1073	2372	3433
51GGC - 350 cm	11375	27.35	9.40	9.43	416	1210	1206
51GGC - 390 cm	6826	15.34	6.05	5.20	445	1129	1312
51GGC - 400 cm	10190	24.10	8.04	7.92	423	1267	1286
12JPC - 55 cm	14637	24.83	10.00	8.21	589	1463	1783
12JPC - 85 cm	12956	40.82	12.79	13.58	317	1013	954
12JPC - 263 cm	13369	39.47	15.41	14.78	339	867	904
				Average:	**495**	**1271**	**1461**

[1] Literature data from Hein et al. (1999)
[2] Where possible Al concentrations were calculated on a loss-on-ignition free base.
[3] Concentration data given for sediments analysed in this study are normalised to μg per gram of raw sediment weighed in.

Table 2.4. Selection of Al, Nd, Pb and Th concentrations for Pacific ferromanganese crusts, as compiled in Hein et al. (1999), and the respective elemental ratios. For comparison, the concentrations and elemental ratios for samples analysed during this study are displayed. Concentrations for the different fractions are given in μg per gram of bulk sediment used. Elemental ratios are also displayed in Figure 2.5.

2.3.5 Elemental ratios

Another independent tool to evaluate the characteristics of an extracted phase from marine sediments is its aluminium content. Compared with crustal rocks or detrital sediments, hydrogenetic ferromanganese crusts generally display low Al/Nd and

41

Al/Pb due to the preferential incorporation of trace metals over Al (see Hein et al. (1999) for a compilation). For comparison we chose data obtained from ferromanganese crusts from the Central equatorial and the Southern Pacific, which are a good approximation for a hydrogenetic Fe-Mn oxyhydroxide end member, as these locations are most remote from any continental source area, representing the most pristine hydrogenetic geochemical composition. In Table 2.4, Al, Nd, Pb and Th concentrations as well as elemental ratios are shown for selected abyssal ferromanganese crusts (data from Hein et al., 1999), and the different sediment fractions analysed in this study are displayed for comparison. The Al/Nd, Al/Pb and Al/Th for the Blake Ridge sediment fractions are plotted in Figure 2.5 (note the log scale of the y-axis). For the first (3 h-leached) Fe-Mn oxyhydroxide fraction, Al/Nd values are even lower than those for ferromanganese crusts in Table 2.4, averaging to ratios of 13.2. The second (24 h-leached) Fe-Mn oxyhydroxide leaches yielded Al/Nd in the range of the presented Fe-Mn crusts (average Al/Nd = 61). The ratios in the detrital fraction of the sediment samples are entirely different from those in the first Fe-Mn oxyhydroxide fraction (average Al/Nd = 495; Table 2.4, Fig. 2.5).

For Al/Pb, the first leaches have comparable elemental ratios to Pacific abyssal ferromanganese crusts, averaging to Al/Pb of 18.3 and 11, respectively (Table 2.4). The second Fe-Mn oxyhydroxide leach fraction has somewhat elevated ratios (Al/Pb = 83), whereas Al/Pb of the remaining detrital fraction are an order of magnitude higher with an average of 1271 (see also Fig. 2.5).

The ratio of Al over Th (^{232}Th), in contrast, reflects the very high particle reactivity of Th compared with Nd or Pb. Ferromanganese crust data displayed in Table 3 have a relatively high Al/Th but this is due to low concentrations of dissolved or particulate ^{232}Th in the deep Pacific (Swarzenski et al., 2003). The continental slope and rise of the eastern North America is an area of high turbidity and high particle fluxes from the continent to the deep Atlantic (Eittreim et al., 1969; Eittreim et al., 1976; McCave, 1986; Biscaye et al., 1988). The waters along this continental margin have highly elevated dissolved ^{232}Th concentrations due to particle-seawater interaction, reflected in very low Al/Th recorded in Fe-Mn oxyhydroxide coatings (Guo et al., 1995). More importantly for our purposes, however, the two sequential leaching phases, as well as the detrital fraction in the sediments show similar systematic differences analogous to

42

Al/Nd and Al/Pb, which are independent of the age of the sediment. Therefore the
very low Al/Nd, Al/Pb and Al/Th in the first Fe-Mn oxyhydroxide leach fraction
support the hydrogenetic origin of all three extracted trace metals.

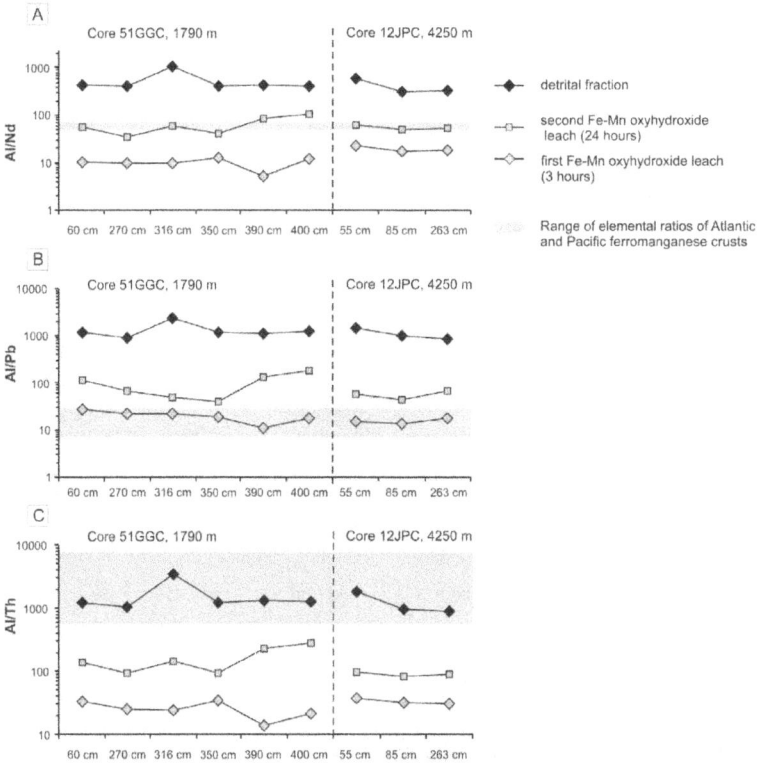

Figure 2.5. Elemental ratios for three different chemical phases of samples presented. Al/Nd for the
leached coatings that were used for the isotope analyses (light grey diamonds) average to 13 and are
lower than hydrogenetic ferromanganese crusts in the equatorial Central and South Pacific (see also
Table 2.4). Al/Pb of the first Fe-Mn oxyhydroxide fraction are similar to Pacific ferromanganese crust
data, and Al/Th are orders of magnitude lower than ferromanganese crust data. This is, however, due to
the absence of dissolved Th in the equatorial Central and South Pacific (see also Table 2.4). Note the
log scale of the y-axis. The individual elemental ratios are also displayed in Table 2.4. The gray boxes
highlight the range of ferromanganese crust elemental ratios shown in Table 2.4.

43

2.4   Discussion

2.4.1   Mass balance calculations

Neodymium and Pb are highly enriched in the Fe-Mn oxyhydroxide fraction of marine sediments compared with the detrital fraction. These elements should thus be less prone to isotopic disturbance by detrital contamination than Sr. In order to constrain the reliability of the extracted ^{87}Sr/^{86}Sr as well as the ^{187}Os/^{188}Os signal as an indicator of detrital contamination of the extracted Nd and Pb signatures we conducted a mass balance test using the elemental and isotopic distributions of Sr, Nd and Pb in the leached Fe-Mn oxyhydroxide coatings and the corresponding detrital fractions.

For the sediment samples the concentrations of Sr, Nd and Pb were measured in both the detrital and the two Fe-Mn oxyhydroxide coating fractions. The Sr isotope composition was determined for the first Fe-Mn oxyhydroxide and the detrital fraction. For simplicity, the second Fe-Mn oxyhydroxide leach and the residual detrital phase are considered together as the detrital phase because this second extracted phase represents a mixture of the authigenic and the detrital fraction. The error introduced into the calculations by considering the second Fe-Mn oxyhydroxide fraction as part of the detrital fraction is considered a conservative buffer because it shifts the individual elemental mass balances towards the detrital fraction. In other words: Detrital contributions to the first extracted authigenic Fe-Mn oxyhydroxide fraction should become even more pronounced, which will be shown to have an insignificant effect for the extracted authigenic Nd and Pb isotope compositions (see below). The present-day seawater ^{87}Sr/^{86}Sr of 0.70918 is used as a reference to determine the detrital contribution f to the Fe-Mn oxyhydroxide leach:

$$f = \frac{(^{87}Sr/^{86}Sr_{mix} - {}^{87}Sr/^{86}Sr_{seawater})}{(^{87}Sr/^{86}Sr_{det\,r} - {}^{87}Sr/^{86}Sr_{seawater})} \quad \text{(Eq. 1)}$$

In Equation 1, ^{87}Sr/^{86}Sr$_{mix}$ represents the measured Sr isotope composition of the first Fe-Mn oxyhydroxide fraction. The total detrital Sr contribution to the first Fe-Mn oxyhydroxide fraction (in ng) is calculated as:

$$[Sr]_{det\,r,coating} = f \times [Sr]_{coating,total} \quad \text{(Eq. 2)}$$

44

The percentage p of Sr leached from the detrital fraction is then determined as

$$p = \left[\frac{[Sr]_{detr,coating}}{([Sr]_{detrital} + [Sr]_{detr,coating})} \right] \times 100 \qquad \text{(Eq. 3)}$$

where p represents the total amount of leached detrital Sr (in %), $[Sr]_{detr, coating}$ is the amount of leached detrital Sr in the coating fraction and $[Sr]_{detrital}$ is the total mass of Sr present in the sample. Assuming that the same percentage of Nd and Pb was leached from the detrital fraction as for Sr, the total detrital Nd and Pb pools can be quantified:

$$[Nd, Pb]_{detr,total} = [Nd, Pb]_{detrital} \times \frac{100}{(100 - p)} \qquad \text{(Eq. 4)}$$

and the amount of Nd and Pb leached from the detrital fraction be calculated:

$$[Nd, Pb]_{detr,coating} = p \times [Nd, Pb]_{detr,total} \qquad \text{(Eq. 5)}$$

$[Nd, Pb]_{detr, coating}$ refers to the amount of leached detrital Nd or Pb in the coating fraction. Following the quantification of the seawater contribution to the Nd and Pb Fe-Mn oxyhydroxide fraction (i.e., $[Nd, Pb]_{sw, coating}$)

$$[Nd, Pb]_{SW,coating} = [Nd, Pb]_{coating,total} - [Nd, Pb]_{detr,coating} \qquad \text{(Eq. 6)}$$

the hypothetical true seawater Nd and Pb isotope composition R can be determined:

$$R_{SW} = \frac{(R_{mix} - g \times R_{detritus})}{(1 - g)} \qquad \text{(Eq. 7)}$$

$R = \varepsilon_{Nd}$, $^{206}Pb/^{204}Pb$, $^{207}Pb/^{204}Pb$, $^{208}Pb/^{204}Pb$, $^{207}Pb/^{206}Pb$, $^{208}Pb/^{206}Pb$. The variable g stands for the detrital contribution to the respective isotope signal (see below). In this case the seawater R was calculated as:

$$R_{SW} = \frac{\left(R_{mix} - \frac{[Nd, Pb]_{detr,coating}}{[Nd, Pb]_{coating,total}} \times R_{detritus} \right)}{\left(\frac{[Nd, Pb]_{SW,coating}}{[Nd, Pb]_{coating,total}} \right)} \qquad \text{(Eq. 8)}$$

Measured Sr, Nd and Pb concentrations, isotope ratios, and individual results of the mass balance calculations

Sample #	$^{87}Sr/^{86}Sr$	Sr [ng/g]	Nd [ng/g]	Pb [ng/g]	f (Eq. 1)	$Sr_{detr, coating}$ [ng/g] (Eq. 2)	p (Eq. 3)	$Nd_{detr, total}$ [ng/g] (Eq. 4)	$Nd_{detr, coating}$ [ng/g] (Eq. 5)	$Nd_{SW, coating}$ [ng/g] (Eq. 6)	$Pb_{detr, total}$ [ng/g] (Eq. 4)	$Pb_{detr, coating}$ [ng/g] (Eq. 5)	$Pb_{SW, coating}$ [ng/g] (Eq. 6)
Core 51GGC -													
60 cm	0.70918	14059	3515	1366	0.00011	2	0.003%	17301	0.5	3515	6413	0.2	1366
60 cm detritus	0.72076	58259	17300	6412									
270 cm	0.70958	2451	4946	2183	0.03019	74	0.085%	26653	22.7	4923	11900	10.1	2173
270 cm detritus	0.72228	86659	26630	11890									
316 cm	0.70949	1083	2495	1140	0.02588	28	0.086%	5979	5.2	2490	2704	2.3	1138
316 cm detritus	0.72108	32383	5973	2701									
350 cm	0.70977	2181	3895	2630	0.03404	74	0.075%	27374	20.6	3874	9410	7.1	2623
350 cm detritus	0.72659	98689	27354	9403									
400 cm	0.70983	1130	2642	1786	0.04715	53	0.054%	24117	13.1	2629	8049	4.4	1782
400 cm detritus	0.72307	98200	24104	8045									
Core 121PC -													
55 cm	0.71500	1368	4189	6346	0.29973	410	0.301%	24905	74.8	4115	10035	30.1	6316
55 cm detritus	0.72861	135697	24830	10005									
85 cm	0.71391	1121	4684	6120	0.22109	248	0.159%	40885	64.9	4619	12809	20.3	6099
85 cm detritus	0.73057	155606	40821	12788									
263 cm	0.71239	870	4248	4317	0.14025	122	0.083%	39499	32.8	4215	15426	12.8	4304
263 cm detritus	0.73207	146781	39466	15413									

Table 2.5. Sr, Nd and Pb concentrations and isotope compositions for the individual samples and the results of the mass balance calculations. Equation numbers in brackets refer to equation numbers presented in the text. Δ refers to the difference between the calculated and the measured seawater isotope compositions.

Sample #	measured ε_{Nd} (detritus)	measured ε_{Nd} (mix)	calculated ε_{Nd} (seawater) (Eq. 8)	Δ sw - mix	measured $^{206}Pb/^{204}Pb$ (detritus)	measured $^{206}Pb/^{204}Pb$ (mix)	calculated $^{206}Pb/^{204}Pb$ (seawater) (Eq. 8)	Δ sw - mix	measured $^{207}Pb/^{204}Pb$ (detritus)	measured $^{207}Pb/^{204}Pb$ (mix)	calculated $^{207}Pb/^{204}Pb$ (seawater) (Eq. 8)	Δ sw - mix
Core 51GGC -												
60 cm	-10.97	-9.97	-9.97	0.0001	18.992	19.134	19.134	0.0000	15.650	15.691	15.691	0.0000
270 cm	-10.83	-10.22	-10.22	0.0028	19.019	19.219	19.220	0.0009	15.649	15.684	15.685	0.0002
316 cm	-11.33	-10.32	-10.32	0.0021	18.882	19.123	19.123	0.0005	15.625	15.678	15.678	0.0001
350 cm	-12.39	-10.72	-10.71	0.0089	18.944	19.072	19.073	0.0003	15.625	15.667	15.667	0.0001
400 cm	-11.81	-9.44	-9.43	0.0118	19.101	18.980	18.980	-0.0003	15.636	15.662	15.662	0.0001
Core 12JPC -												
55 cm	-15.02	-13.17	-13.14	0.0336	18.32	19.377	19.382	0.0051	15.527	15.693	15.694	0.0008
85 cm	-13.56	-11.10	-11.06	0.0346	18.69	19.112	19.113	0.0014	15.589	15.665	15.665	0.0003
263 cm	-12.21	-10.11	-10.10	0.0163	18.87	18.944	18.944	0.0002	15.625	15.654	15.654	0.0001

Sample #	measured $^{208}Pb/^{204}Pb$ (detritus)	measured $^{208}Pb/^{204}Pb$ (mix)	calculated $^{208}Pb/^{204}Pb$ (seawater) (Eq. 8)	Δ sw - mix	measured $^{207}Pb/^{206}Pb$ (detritus)	measured $^{207}Pb/^{206}Pb$ (mix)	calculated $^{207}Pb/^{206}Pb$ (seawater) (Eq. 8)	Δ sw - mix	measured $^{208}Pb/^{206}Pb$ (detritus)	measured $^{208}Pb/^{206}Pb$ (mix)	calculated $^{208}Pb/^{206}Pb$ (seawater) (Eq. 8)	Δ sw - mix
Core 51GGC -												
60 cm	38.886	39.162	39.162	0.0000	0.8240	0.8201	0.8201	0.00000	2.0475	2.0468	2.0468	0.00000
270 cm	38.902	39.274	39.275	0.0017	0.8228	0.8161	0.8161	-0.00003	2.0455	2.0435	2.0435	-0.00001
316 cm	38.739	39.131	39.131	0.0008	0.8275	0.8199	0.8199	-0.00002	2.0516	2.0463	2.0463	-0.00001
350 cm	38.874	39.076	39.076	0.0005	0.8248	0.8215	0.8215	-0.00001	2.0521	2.0488	2.0488	-0.00001
400 cm	38.969	38.898	38.897	-0.0002	0.8186	0.8252	0.8252	0.00002	2.0401	2.0494	2.0494	0.00002
Core 12JPC -												
55 cm	38.522	39.591	39.596	0.0051	0.8477	0.8099	0.8097	-0.00018	2.1033	2.0432	2.0429	-0.00029
85 cm	38.804	39.173	39.174	0.0012	0.8340	0.8197	0.8196	-0.00005	2.0759	2.0497	2.0496	-0.00009
263 cm	38.935	38.921	38.921	0.0000	0.8281	0.8264	0.8264	-0.00001	2.0635	2.0545	2.0545	-0.00003

Table 2.5. *continued.*

Table 2.5 summarises the results for the individual calculations. The results of these mass balance calculations provide compelling evidence that the $^{87}Sr/^{86}Sr$ signal of the Fe-Mn oxyhydroxide fraction does not yield reliable information on the seawater origin of the measured Nd and Pb isotope signal. Not a single calculated seawater Nd and Pb isotope composition listed in Table 2.5 differs significantly from the measured Fe-Mn oxyhydroxide composition. For Nd the calculated offset from the true seawater signal is always significantly smaller than the analytical 2σ uncertainty of ±0.27 ϵ_{Nd}. Indeed, even for the highest measured $^{87}Sr/^{86}Sr$, the offset of the calculated Nd seawater composition is an order of magnitude smaller than the external reproducibility (sample JPC12 – 55 cm, $^{87}Sr/^{86}Sr$ = 0.71500; Table 2.5). A similar situation is observed for the Pb isotope signal. The offset of the calculated true seawater signal is always on the order of or smaller than the external reproducibility. Because of the large variability observed for the seawater Pb isotope signal on glacial-interglacial timescales, this renders the Pb isotope signal even more reliable than the Nd isotope signal. These calculations clearly imply that the present-day seawater $^{87}Sr/^{86}Sr$ can very well serve as proof for the seawater origin of the extracted Fe-Mn oxyhydroxide phase, such as shown for sites in the Cape Basin (cf., Piotrowski et al., 2005), but it cannot serve to disprove the seawater origin of the extracted Nd and Pb fraction.

The robustness of the extracted Fe-Mn oxyhydroxide Nd and Pb isotope signals can be tested further for hypothetical extreme Nd and Pb isotope composition of the detrital fraction. Figure 2.6a illustrates the effect of altering the Nd isotope composition of the detrital fraction in the sediment. Even for unrealistically low $\epsilon_{Nd, detritus}$ of –25 and a measured $^{87}Sr/^{86}Sr$ of 0.71209, the calculated seawater ϵ_{Nd} is offset from the measured value only by –0.10 ϵ_{Nd} units. In every case presented in Figure 2.6a the offset of the calculated ϵ_{Nd} from the measured ϵ_{Nd} was significantly smaller than the long-term analytical external reproducibility of ± 0.27 ϵ_{Nd}. Equally consistent calculated seawater Pb isotope compositions are found when modifying the detrital Pb isotope composition. In Figure 2.6b the Pb isotopic behaviour is illustrated for $^{206}Pb/^{204}Pb$. Changing the $^{206}Pb/^{204}Pb$ of the detrital phase to ratios as low as 16.0 only modifies the calculated seawater Pb isotope composition insignificantly (Fig. 2.6b). The other Pb isotope ratios provide identical information and are not displayed.

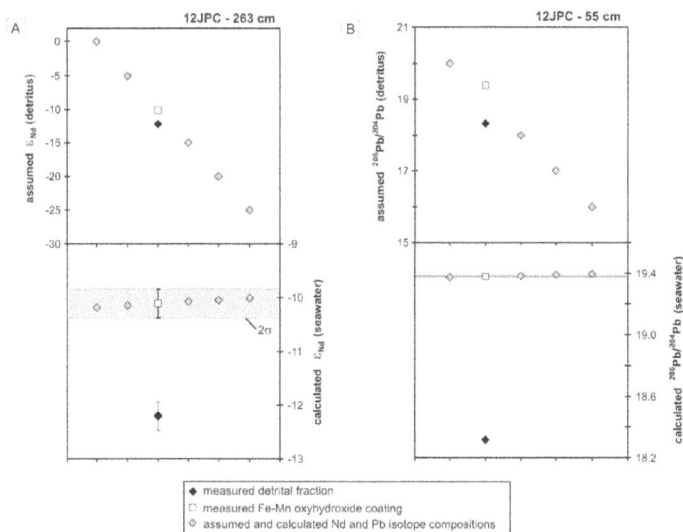

Figure 2.6.   (A) Calculated true ε_{Nd} for the extracted Fe-Mn oxyhydroxide fraction (lower panel), assuming a variety of possible detrital Nd isotope compositions (upper panel). Even detrital ε_{Nd} as low as -25 and as high as 0 do not change the calculated true seawater ε_{Nd} beyond the external reproducibility of 0.27 ε_{Nd}. (B) Calculated true $^{206}Pb/^{204}Pb$ for the extracted Fe-Mn oxyhydroxide fraction (lower panel), assuming a variety of possible complementary detrital $^{206}Pb/^{204}Pb$ isotope compositions (upper panel). Even detrital ε_{Nd} as low as 16.0 does not change the calculated true seawater $^{206}Pb/^{204}Pb$ to any significant extent.

2.4.2   Thorium isotopes in Fe-Mn oxyhydroxide coatings

Due to the particle reactivity of Th $^{232}Th/^{230}Th$ should be low in seawater-derived Fe-Mn oxyhydroxide coatings, as ^{232}Th can only be supplied to the oceans via continental input and is quickly removed at ocean margins (Guo et al., 1995; Swarzenski et al., 2003), whereas ^{230}Th is continuously produced in the water column through the decay of ^{238}U and ^{234}U. Uranium behaves conservatively in seawater under oxic conditions with a residence time of several 100 kyrs, which is the reason why the production rate of ^{230}Th is homogeneous in all ocean basins (Nozaki, 1991; Dunk et al., 2002). Earlier studies demonstrated a systematic change in $^{232}Th/^{230}Th$ from proximal continental to distal marine environments (Guo et al., 1995; Robinson et al., 2004). Robinson et al. (2004) presented seawater $^{232}Th/^{230}Th$ in the range of 13,000 to 31,600 in depths between the surface and 1,650 m water depth in both waters and carbonates close to the Bahamas. Although this location is relatively close to the Blake Ridge, it is isolated from direct continental influences because of the prevailing circulation

49

regime (Anselmetti et al., 2000). Considering the proximity of the Blake Ridge to the continental rise of eastern North America, the Holocene $^{232}Th/^{230}Th$ compositions seem to reflect a pure hydrogenetic signal, unaffected by detrital contributions. On the other hand, the ratios are significantly higher during the LGM and the deglaciation. The reasons for this can be twofold: Either sediment deposited during this interval was more prone to contemporaneous leaching of detrital Th, or the $^{232}Th/^{230}Th$ isotope composition along the Blake Ridge varied during the transition from the LGM to the Holocene.

It seems unlikely that partial leaching of the detrital phase is the cause for the higher ratios of Fe-Mn oxyhydroxide fractions older than the Holocene. Figure 2.2b illustrates that there is no obvious correlation between $^{87}Sr/^{86}Sr$ and $^{232}Th/^{230}Th$, arguing against a leaching effect as the cause for the variation seen in $^{232}Th/^{230}Th$ in the Fe-Mn oxyhydroxide fractions. More importantly, in Figure 2.5 Al/Th remains systematically low independent of the age of the sediment. Guo et al. (1995) published dissolved and particulate Th isotope compositions from waters offshore Cape Hatteras to the North of the Blake Ridge, ranging from 122,000 in surface water samples to 28,000 in 750 m water depth for dissolved $^{232}Th/^{230}Th$. Particulate $^{232}Th/^{230}Th$ was significantly higher, ranging from 231,000 to 143,000 in the same depth range.

Analogous to Nd and Pb, Th concentrations of the different sediment fractions shown in Table 2.4 indicate that a significant amount of bulk Th is locked in the seawater-derived fraction in sediments along the Blake Ridge, hence the differences in Al/Th between the three different fractions displayed in Table 2.4 and Figure 2.5 are even more extreme than for Al/Nd and Al/Pb. The systematic variation between the two sequentially leached Fe-Mn oxyhydroxide fractions and the detrital fraction, which are independent of the age and the depth of the sediment cores, strongly suggests that the observed Th isotope signal is of seawater origin. The observation that $^{234}Th/^{232}Th$ in unfiltered water and particles are similar (Swarzenski et al., 2003) supports the notion that Th isotopes are not fractionated during incorporation into Fe-Mn oxyhydroxide coatings.

The amplitude of variation observed in $^{232}Th/^{230}Th$ between the LGM and the Holocene is not surprising considering the very short Th residence time in seawater. The residence time of dissolved ^{230}Th is on the order of <4 years in ocean margin settings and as long as 40 years in central ocean deep waters (Anderson et al., 1983a; Anderson et al., 1983b; Guo et al., 1995), which makes short-term seawater $^{232}Th/^{230}Th$ isotope variations very likely. The highest $^{232}Th/^{230}Th$ in core 51GGC was recorded at about 17.2 ka BP, immediately after the LGM. Considering the age uncertainties associated with radiocarbon dating during the Deglaciation (Stuiver et al., 1998; Hughen et al., 2004a) and the error induced by the uncertainty of the lock-in depth of the authigenic Fe-Mn oxyhydroxide signal (*i.e.* depth in the sediment below which the early diagenetic Fe-Mn oxyhydroxide is not modified anymore), the $^{232}Th/^{230}Th$ may have recorded the highly elevated IRD particle flux in the North Atlantic associated the large scale iceberg discharge event Heinrich 1 (~16.8 ka BP; Hemming, 2004).

2.4.3   Rare earth element patterns

The REE patterns shown in Figure 2.4 are similar to anoxic marine pore water REE patterns (Haley et al., 2004). These authors attributed the MREE enrichment to the release of REE from Fe oxide phases under reducing conditions, which preferentially release the incorporated MREE. Conversely, Hannigan and Sholkovitz (2001) reported MREE enrichments in waters from a variety of rivers being tightly coupled with dissolved phosphate concentrations. Hence the MREE enrichment could be caused by two different phases, either by leaching of phosphates or by release of REE from the Fe-Mn oxyhydroxides. Ohr et al. (1994) argued that the REE budget obtained during leaching of argillaceous sediments may be controlled by dissolution of early diagenetic apatite. Contrary to that, Haley et al. (2004) found no unambiguous indication that the MREE enrichment in porewaters containing dissolved phosphate is controlled by phosphates such as authigenic apatite. We cannot fully rule out that part of the REE signal extracted during leaching of the Fe-Mn oxyhydroxide fraction could be derived from contemporaneous leaching of authigenic apatite, but because of the preceding carbonate removal step this seems unlikely: Authigenic apatite can be easily dissolved using dilute acetic acid (Ohr et al., 1994), and the very first chemical reagent used here for the extraction of the carbonate fraction, preceding the reductive leaching of Fe-Mn oxyhydroxides, is a Na acetate buffer containing acetic acid (Fig. 2.1). Therefore, we attribute the MREE-enriched patterns shown in Figure 2.4a and 2.4b to the release of REE from Fe-Mn oxyhydroxide coatings.

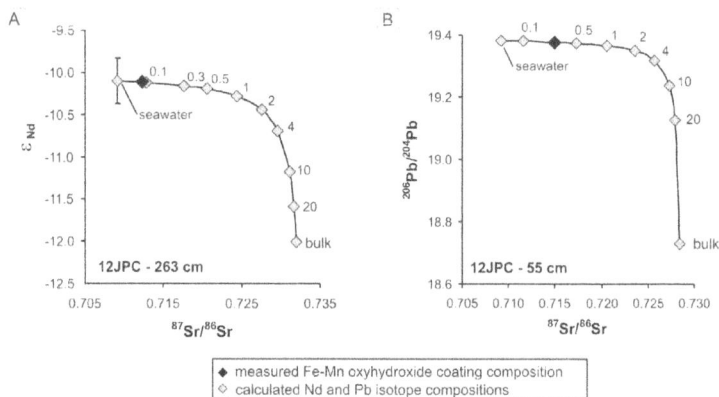

Figure 2.7. (A) Measured and calculated ε_{Nd} for the extracted Fe-Mn oxyhydroxide fraction with increasing contemporaneous leaching of the detrital phase, exemplified for sample 12JPC – 263 cm. Numbers refer to the percentage of the detrital fraction that is extracted together with the Fe-Mn oxyhydroxide fraction. The Figure demonstrates why even contributions as low as 0.1 % of the detrital fraction already significantly disturb the Sr isotope composition, which, however, has no effect for the Nd isotope composition. Calculated ratios are also displayed in Table 2.6. (B) Measured and calculated $^{206}Pb/^{204}Pb$ for the extracted Fe-Mn oxyhydroxide fraction analogous to A, shown for sample 12JPC – 55 cm. Similar to the Nd isotope composition of the Fe-Mn oxyhydroxide fraction, the Pb isotope composition is very robust against detrital contamination if the Pb isotope system. The remaining Pb isotope compositions follow identical detrital phase admixing paths, which is why they are not shown here. Calculated Pb ratios are also displayed in Table 2.7.

2.4.4 Neodymium and lead mass balance

The reason for the consistency of the extracted Nd and Pb isotope signal in the mass balance calculations can best be illustrated by determining the induced offset from seawater compositions by the admixture of a detrital Sr, Nd and Pb. Knowing the total amount of seawater-derived Sr, Nd and Pb from Equation 6, as well as the concentrations and the isotope compositions of the respective fractions, the imprint of detrital contamination to the seawater isotope signal can be determined. This is done in Tables 2.6, 2.7 and Figure 2.7, which highlight the susceptibility of the Sr isotope signal of the Fe-Mn oxyhydroxide fraction to detrital contamination, accompanied by only insignificant offsets for the extracted Nd and Pb signal of the same fraction. The numbers along the mixing curve in Figure 7 refer to the percentage of the detrital fraction that is extracted together with the Fe-Mn oxyhydroxide pool.

Even if 1 % of the entire detrital fraction was extracted with the seawater-derived fraction, for mass balance constraints the Nd and Pb isotope signal would still be dominated by the Fe-Mn oxyhydroxide pool. In fact, all of the sediment samples processed for the mass balance calculation have 0.3 % or less detrital contribution (p in Table 2.5).

Neodymium mixing relationships exemplified with sample JPC12 - 263 cm.

measured

	$^{87}Sr/^{86}Sr$	ε_{Nd}		Sr [ng/g]	Nd [ng/g]
12JPC - 263 cm coating	0.71239	-10.11		870	4248
12JPC - 263 cm detritus	0.73207	-12.21		146781	39466

calculated

	Nd [ng/g]
seawater Nd	4215
detrital Nd	39499
calculated bulk Nd	43713

	Sr [ng/g]	$^{87}Sr/^{86}Sr$
seawater Sr	748	0.70918
detrital Sr	146903	0.73207
calculated bulk Sr	147652	0.73195

detrital contribution to leachate (%)	detrital Sr in leachate [ng/g]	total Sr in leachate [ng/g]	calculated $^{87}Sr/^{86}Sr$	detrital Nd in leachate [ng/g]	total Nd in leachate [ng/g]	calculated ε_{Nd}
0.0%	0	748	0.70918	0	4215	-10.10
0.1%	147	894	0.71294	39	4250	-10.12
0.3%	441	1187	0.71768	118	4321	-10.16
0.5%	735	1479	0.72055	197	4391	-10.20
1.0%	1469	2210	0.72440	395	4568	-10.28
2.0%	2938	3671	0.72750	790	4920	-10.44
4.0%	5876	6594	0.72958	1580	5626	-10.69
10.0%	14690	15364	0.73107	3950	7743	-11.18
20.0%	29381	29979	0.73161	7900	11272	-11.58
bulk	146903	147652	0.73195	39499	43713	-12.01

Table 2.6. Coupled Sr and Nd isotope mixing relationships, as shown in Figure 2.7a, for the Fe-Mn oxyhydroxide fraction, assuming different proportions of admixture of the detrital component, illustrated for sample 12JPC – 263 cm.

Lead mixing relationships exemplified with sample JPC12 - 55 cm.

measured

	$^{87}Sr/^{86}Sr$	$^{206}Pb/^{204}Pb$	$^{207}Pb/^{204}Pb$	$^{208}Pb/^{204}Pb$	$^{207}Pb/^{206}Pb$	$^{208}Pb/^{206}Pb$	Sr [ng/g]	Pb [ng/g]
12JPC - 55cm coating	0.71500	19.377	15.693	39.591	0.810	2.043	1368	6346
12JPC - 55cm detritus	0.72861	18.316	15.527	38.522	0.848	2.103	135697	10005

calculated

	$^{206}Pb/^{204}Pb$	$^{207}Pb/^{204}Pb$	$^{208}Pb/^{204}Pb$	$^{207}Pb/^{206}Pb$	$^{208}Pb/^{206}Pb$	Pb [ng/g]
seawater Pb	19.382	15.694	39.596	0.810	2.043	6316
detrital Pb	18.316	15.527	38.522	0.848	2.103	10035
calculated bulk Pb	18.728	15.591	38.937	0.833	2.080	16351

	$^{87}Sr/^{86}Sr$	Sr [ng/g]
seawater Sr	0.70918	958
detrital Sr	0.72861	136107
calculated bulk Sr	0.72847	137065

detrital contribution to leachate (%)	detrital Sr in leachate [ng/g]	total Sr in leachate [ng/g]	calculated $^{87}Sr/^{86}Sr$	detrital Pb in leachate [ng/g]	total Pb in leachate [ng/g]	calculated $^{206}Pb/^{204}Pb$	calculated $^{207}Pb/^{204}Pb$	calculated $^{208}Pb/^{204}Pb$	calculated $^{207}Pb/^{206}Pb$	calculated $^{208}Pb/^{206}Pb$
0.00%	0	958	0.70918	0	6316	19.382	15.694	39.596	0.8097	2.0429
0.1%	136	1094	0.71160	10	6326	19.381	15.693	39.594	0.8097	2.0430
0.3%	408	1366	0.71499	30	6346	19.377	15.693	39.591	0.8099	2.0432
0.5%	681	1639	0.71725	50	6366	19.374	15.692	39.587	0.8100	2.0434
1.0%	1361	2319	0.72058	100	6416	19.366	15.691	39.579	0.8103	2.0438
2.0%	2722	3680	0.72355	201	6517	19.349	15.688	39.563	0.8109	2.0448
4.0%	5444	6402	0.72570	401	6718	19.319	15.684	39.532	0.8120	2.0465
10.0%	13611	14569	0.72733	1003	7320	19.236	15.671	39.449	0.8149	2.0512
20.0%	27221	28180	0.72795	2007	8323	19.125	15.653	39.337	0.8189	2.0574
bulk	137065	137065	0.72847	10035	16351	18.728	15.591	38.937	0.8330	2.0799

Table 2.7. Coupled Sr and Pb isotope mixing relationships, as shown in Figure 2.7b, for the Fe-Mn oxyhydroxide fraction, assuming different proportions of admixture of the detrital component, illustrated for sample 12JPC 55 cm.

54

2.5   Conclusions

The seawater-derived Fe-Mn oxyhydroxide fraction, extracted from marine drift sediments in the North Atlantic, is characterised by means of geochemical, radiogenic isotopic, and mass balance approaches to constrain the purity of the recorded seawater signal. Although the Sr and Os isotope compositions of the leached Fe-Mn oxyhydroxide fractions suggest contributions from the detrital pool in the sediments for these elements, this contribution is insignificant for Nd, Pb and Th isotopes due to mass balance constraints.

Thorium isotopes show a marked variation on glacial-interglacial timescales. Our data suggest that the seawater $^{232}Th/^{230}Th$ dropped significantly during the transition from the LGM to the Holocene. The high $^{232}Th/^{230}Th$ might be attributable to an increased particle load of the water masses in the western North Atlantic during the LGM, Heinrich 1 and the early deglaciation.

Al/Nd, Al/Pb and Al/Th ratios reveal a strong preferential enrichment of the trace metals relative to Al in the leached Fe-Mn oxyhydroxide phase compared with the detrital fraction. These elemental ratios are on the order of or lower than those for abyssal ferromanganese crusts from the central and South Pacific. The large difference in Al/Nd, Al/Pb and Al/Th ratios between the Fe-Mn oxyhydroxide fraction and the residual detrital fraction reflects the hydrogenous origin of the leached Fe-Mn oxyhydroxides.

The Fe-Mn oxyhydroxide fraction that was extracted from pelagic sediments displays REE patterns virtually identical to Fe-rich anoxic marine pore waters, in which the dissolved trace metals are ultimately of seawater orgin. The observed MREE enrichment is controlled by REE released from Fe-Mn oxyhydroxides. Although authigenic apatites would create similar MREE-enriched PAAS-normalised REE patterns their contribution to the leached Fe-Mn oxyhydroxide signal is regarded as insignificant because of the preceding chemical treatment of the samples with buffered acetic acid.

The mass balance calculations demonstrate that the Nd and Pb isotope signature of the leached Fe-Mn oxyhydroxide fraction is not sensitive to detrital contamination, even for extremely elevated $^{87}Sr/^{86}Sr$ or unrealistically low detrital ε_{Nd}. This observation is ascribed to the strong enrichment of Nd and Pb in the authigenic fraction of the sediments.

Elemental ratios such as Al/Nd, Al/Pb or Al/Th, as well as REE patterns contain valuable information regarding the assessment of whether the extracted authigenic Fe-Mn oxyhydroxide fraction was contaminated through detrital contributions. Carrying out these tests for drift sediments along the Blake Ridge confirmed that a pure seawater component for Nd and Pb can be extracted. This applies to marine sediments analysed from this specific location. Although we expect sediments in different locations in the oceans to show similar trace metal distributions, this has to be confirmed on a site-specific basis.

Acknowledgments

M. B. Andersen, E.-K. Potter and M. Wipf are acknowledged for providing assistance with the Th separation and measurements. Dieter Garbe-Schönberg at the Geological Institute of the University of Kiel carried out the major and trace element ICP-AES and ICP-MS. Brian Haley and Sidney Hemming provided additional constructive suggestions.

Chapter 3[*]

Neodymium isotope evolution of North Atlantic Deep and Intermediate Waters in the western North Atlantic since the Last Glacial Maximum

[*]submitted to *Earth and Planetary Science Letters* as: Gutjahr, M. H., Frank, M., Stirling, C. H., Keigwin, L. D. and Halliday, A. N. Neodymium isotope evolution of North Atlantic Deep and Intermediate Waters in the western North Atlantic since the Last Glacial Maximum.

Abstract

A high-resolution seawater Nd isotope record has been extracted from the Fe-Mn oxyhydroxide fraction of pelagic sediments along the Blake Ridge in the North Atlantic in order to ascertain Nd isotope variations between the water masses and to reconstruct the timing of major hydrographic changes in the western North Atlantic since the Last Glacial Maximum (LGM). The Blake Ridge is ideally suited for this because it is located in the major flow path of the Deep Western Boundary Current (DWBC), which transports North Atlantic Deep Water (NADW) southward.

A typical NADW ε_{Nd} value of -13.5 was extracted from sediments located within the main water body of the DWBC below 3200 m water depth. Above this depth the extracted seawater Nd isotopic composition appears to reflect a contribution from surface waters, most likely caused by downslope redistribution of shallower North American shelf sediments within intermediate nepheloid layers. Highly elevated Holocene $^{230}Th_{xs}$ fluxes determined in sediment core 51GGC from 1790 m water depth provide evidence of such an export of sediment from the North American shelf.

The unradiogenic Nd isotopic composition typical of present day NADW is not detectable along the Blake Ridge at all for samples corresponding to the LGM. Contrary to the sections defining the deglaciation and the Holocene, the core from 1790 m water depth did not experience significant sediment focussing during the LGM, as evidenced by the $^{230}Th_{xs}$ data. Therefore the proposed mechanism behind the export of a surface seawater Nd isotope signal into intermediate depths seems to be de-activated, strongly suggesting that the Nd isotope signal at this site during the LGM reflects ambient seawater. Our results indicate that the ε_{Nd} of the shallower glacial equivalent of NADW, the Glacial North Atlantic Intermediate Water (GNAIW) was -9.7 ± 0.4 and thus significantly different from interglacial Upper NADW. Such a radiogenic ε_{Nd} suggests that GNAIW did not contain large amounts of water that passed through the Labrador Sea prior to its arrival at the Blake Ridge. Core locations below 3400 m water depth were located within Southern Source Waters with an ε_{Nd} of -10.4 ± 0.7 during the LGM and the deglaciation. The change to Nd isotopic compositions reflecting a modern circulation pattern, including the presence of Lower NADW, only occurred after the Younger Dryas.

3.1   Introduction

Considerable research efforts have been invested to reconstruct the past variability of the rate of meridional overturning circulation. A variety of proxies have been applied to better characterise past water mass distributions during the transition from the LGM to the Holocene, mostly based on nutrients ($\delta^{13}C$, Cd/Ca) (Boyle and Keigwin, 1987; Duplessy et al., 1988; Sarnthein et al., 1994; Marchitto et al., 2002; Curry and Oppo, 2005), oxygen isotopes of planktonic and benthic foraminifera (Labeyrie et al., 1992), water mass ^{14}C ventilation ages (Keigwin, 2004; Robinson et al., 2005), or ratios of water-borne radioactive trace metals such as $^{231}Pa/^{230}Th$ (McManus et al., 2004; Gherardi et al., 2005). Despite these efforts there are discrepancies regarding the evolution of the water mass structure in the North Atlantic as well as the timing of circulation changes since the Last Glacial Maximum (LGM). For example, it is currently not resolved whether modes of ocean circulation in the North Atlantic were the same during the Younger Dryas and the LGM (e.g., Duplessy et al., 1988; Labeyrie et al., 1992; Keigwin, 2004).

Today the Deep Western Boundary current (DWBC), which transports North Atlantic Deep Water (NADW) southward, is a vertically stratified mixture of water masses derived from various regions in the North Atlantic basin. The Lower NADW (LNADW) is dominated by waters originally derived from the NE Atlantic, Denmark Strait and Subpolar Mode Water, whereas the Upper NADW (UNADW) in the Western North Atlantic mainly consists of water derived from the Labrador Sea and Subpolar Mode Water (SPMW) (McCartney and Talley, 1982; Talley and McCartney, 1982; Schmitz and McCartney, 1993; Dickson and Brown, 1994). The presence of Antarctic Bottom Water (AABW) can be identified both in the deep eastern and western North Atlantic (Speer and McCartney, 1992; Schmitz and McCartney, 1993). There is consensus that Glacial North Atlantic Intermediate Water (GNAIW) replaced NADW during the LGM (Boyle and Keigwin, 1987; Keigwin, 2004; Curry and Oppo, 2005; Robinson et al., 2005). However, it is not clear whether GNAIW simply lacked the LNADW component or if the origin and flow path of the GNAIW was entirely different from the Holocene situation. The subduction of the GNAIW was shallower and water depths below 2000 to 2500 m were most likely dominated by northward advection of Southern Source Water (SSW). It is, however, not well constrained when the major hydrographic changes and the switch to modern circulation patterns occurred. For example, McManus et al. (2004) inferred a near total collapse of the meridional overturning circulation (MOC) contemporaneous with catastrophic iceberg discharge event H1 followed by rapid MOC resumption from the ^{231}Pa/^{230}Th record of a sediment core on the Bermuda Rise in the North Atlantic. Furthermore, their data suggest that the MOC at its present-day mode of vigour was not attained before the beginning of the Holocene. Robinson et al. (2005) suggested brief episodes of deep ventilation in an otherwise dominantly shallowly-ventilated North Atlantic during the deglaciation. These authors did not find evidence for a complete shutdown of the MOC during H1 or any other time since the LGM. The radiocarbon record from deep-sea corals and paired benthic planktonic foraminifera presented by Robinson et al. (2005) also indicated the presence of a radiocarbon-depleted intermediate depth water mass during H1 and the Younger Dryas akin to modern Antarctic Intermediate Water (AAIW). Such a finding was also reported by Rickaby and Elderfield (2005) using Cd/Ca in conjunction with δ^{13}C in benthic foraminifera in a sediment core from the central North Atlantic south of Iceland. Their record indicated the presence of a nutrient-enriched intermediate depth water mass akin to AAIW during H1 and the Younger Dryas, a hydrographic situation completely different from today. Hence, while reconstructions of the hydrographic situation during the LGM and the Holocene are less controversial, those spanning the deglaciation are.

It has been demonstrated that Nd isotopes are a powerful proxy for past water mass structure and mixing, which led to inference of major circulation changes and weathering inputs (Palmer and Elderfield, 1985b; Reynolds et al., 1999; Vance and Burton, 1999; Frank et al., 2002; Scher and Martin, 2004). Neodymium has a residence time in the ocean of 600-2,000 years (Frank, 2002), which makes its isotopic composition a suitable quasi-conservative water mass tracer in the Atlantic Ocean in view of water mass residence times of only few 100 years. Biological fractionation processes do not affect Nd isotopes. ^{143}Nd is the decay product of ^{147}Sm ($T_{1/2} = 106$ Ga), and the dissolved ^{143}Nd/^{144}Nd in the ocean is a function of the flow path of a water mass (Frank, 2002; Goldstein and Hemming, 2003). In Precambrian crustal provinces the inherited Nd is very unradiogenic (low ^{143}Nd/^{144}Nd), whereas young mantle-derived rocks display Nd isotope compositions that are highly radiogenic relative to the average Earth composition. Neodymium is dominantly supplied to the oceans via weathering of the continental crust through rivers, sub-glacial meltwater or leaching of shelf sediments to the adjacent ocean basins, and can only be subsequently modified by water mass mixing. The Nd isotope composition of dissolved Nd in river water generally matches that of the bulk source area (Goldstein and Jacobsen, 1987; Dahlqvist et al., 2005). Slight offsets have only been observed in glaciated terrains due to preferential weathering of certain mineral phases (Andersson et al., 2001; von Blanckenburg and Nagler, 2001). Particle-seawater interaction has also been suggested to be capable of modifying the Nd isotope composition of water masses (Tachikawa et al., 1999; Lacan and Jeandel, 2005a). For convenience the $^{143}Nd/^{144}Nd$ composition of a sample is generally normalised to the Chondrite Uniform Reservoir (CHUR) in epsilon notation:

$$\varepsilon_{Nd} = \left[\frac{^{143}Nd/^{144}Nd_{sample}}{^{143}Nd/^{144}Nd_{CHUR}} - 1 \right] \times 10^{4}$$

($^{143}Nd/^{144}Nd_{CHUR}$ = 0.512638 (Jacobsen and Wasserburg, 1980)).

The Nd isotopic composition provides a particularly powerful tracer in the North Atlantic because through admixtures from the Labrador Sea NADW receives extremely unradiogenic Nd isotope compositions from weathering of Archean and Proterozoic continental crust in northern Canada and Greenland (Piepgras and Wasserburg, 1987; Lacan and Jeandel, 2005a). Lacan and Jeandel (Lacan and Jeandel, 2005a) suggested that, as the various proto-NADW water masses pass the Flemish Cape off Newfoundland, the NADW ε_{Nd} is already fairly homogeneous with an overall vertical variability of 1.3 ε_{Nd} (-14.5 ≤ ε_{Nd} ≤ -13.2). Earlier work by Piepgras and Wasserburg (1987) indicated that in the vicinity of the Blake Ridge the ε_{Nd} of the Deep Western Boundary Current (DWBC), which transports NADW southward, is fully homogenised at a value of -13.5 ± 0.5. Other water masses in the Atlantic have more radiogenic Nd isotope compositions. Intermediate and deep waters in the Iceland, Norwegian and Greenland basins display ε_{Nd} on the order of -7.7 to -10.7 (Lacan and Jeandel, 2004b), Mediterranean outflow water supplies an ε_{Nd} signature between -9.4 and -10.1 to the eastern Atlantic (Tachikawa et al., 2004), and Antarctic Intermediate and Bottom Waters in the Southern Atlantic presently range between -7 and -9 in ε_{Nd} .

In studies carried out in the Cape Basin in the South Atlantic Nd isotopes extracted from Fe-Mn oxyhydroxides in pelagic sediments were successfully applied to monitor changes in the isotopic composition of Southern Ocean water masses, which were interpreted as changes in the strength of the export of NADW to the Southern Ocean on glacial-interglacial and millennial timescales (Jeandel, 1993; Rutberg et al., 2000; Piotrowski et al., 2004; Piotrowski et al., 2005). It was concluded from positive shifts of up to 3 ε_{Nd} units that the export of NADW into the Cape Basin was significantly reduced during the last glacial maximum and during the stadials of the last glacial stage between 60,000 and 20,000 years ago. However, these conclusions rely on the assumption that the Nd isotope composition of NADW and GNAIW has remained constant over the past 100 kyr. Recent studies provide support for this assumption (Foster and Vance, 2006; van de Flierdt, 2006; Foster et al., 2007), but none of these provide constraints on short-term variations during extreme climatic conditions such as the LGM. Given the suggestion of a shutdown of deep water formation in the Labrador Sea during the LGM, the GNAIW Nd isotope composition during Marine Isotope Stage 2 (MIS 2) was likely different from the GNAIW and NADW during MIS 1 and 3 if changes in water mass contributions indeed occurred (de Vernal et al., 2002; Cottet-Puinel et al., 2004).

The aim of the present study is to reconstruct the Nd isotope evolution of NADW and GNAIW close to its site of formation between the LGM and today. We focus on the reconstruction of the water mass structure and the timing of hydrographic changes above the Blake Ridge in the western subtropical North Atlantic. This location is ideal to detect past changes in southward water mass export because its sediments are located within the main flow path of the DWBC on the continental rise of eastern North America (Fig. 3.1). In addition, $^{230}Th_{xs}$ analyses were carried out in a sediment core in intermediate depth at the Blake Ridge in order to monitor possible sediment re-distribution phenomena, a crucial parameter in a study like this that focuses on a sediment drift deposit.

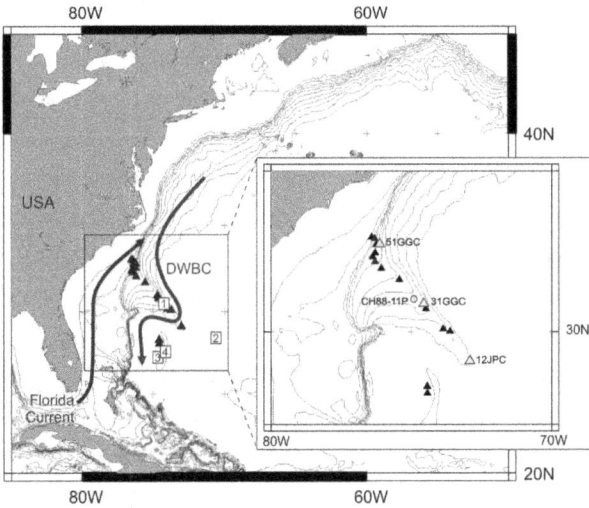

Figure 3.1.   Study area along the Blake Ridge in the western North Atlantic. The Deep Western Boundary Current passes the Blake Ridge southward centred between about 3500 and 4200 m water depth (Stahr and Sanford, 1999). The Florida Current transports Gulf Stream water on the Blake Plateau to the west of the Blake Ridge at water depths above about 700 m water depth (Lynch-Stieglitz et al., 1999). Black triangles represent core locations chosen for either core-top or LGM analyses. The three highlighted core sites represent locations for which Nd downcore analyses were carried out. Core location of CH88-11P studied by (Luo et al., 2001) is also shown. Numbers in white boxes refer to sampling stations for seawater Nd isotope profile OCE63 (Piepgras and Wasserburg, 1987) plotted in Figure 3.3 (Station 1: 300 m; Station 2: 1000, 2200, 3400 m; Station 3: 50 m; Station 4: 4100 m water depth).

3.2   Material and Methods

3.2.1   Nd isotopes

In order to document the present-day Nd isotope composition of water masses along the Blake Ridge, 22 core-top sediment samples from water depths between 775 m and 4712 m, recovered during R/V Knorr cruise 140 (KNR140) were analysed (Fig. 3.1). Eight cores from the KNR140 cruise were selected for analyses of samples from the LGM. At three sites (51GGC, 1780 m; 31GGC, 3410 m; 12JPC, 4250 m) further downcore analyses were carried out. Additionally, the LGM sequence of ODP Leg 172, Site 1054A in 1300 m water depth was analysed. In general, bulk sediments were processed with the exception of the core-top samples, for which only the coarse fraction >63 µm was used. The extracted Fe-Mn oxyhydroxide Nd isotope signal from the bulk and the coarse fraction was found to be identical.

63

The method applied for the extraction of seawater Nd isotopic compositions from Fe-Mn oxyhydroxide coatings was modified from a selection of existing sequential extraction methods (Chester and Hughes, 1967; Tessier et al., 1979; Rutberg et al., 2000; Bayon et al., 2002). Carbonate was removed using a Na acetate buffer followed by removal of adsorbed metals using a 1M $MgCl_2$ solution. Following three rinses in deionised water (Milli-Q system), the oxyhydroxide coatings were leached using a 0.05M hydroxylamine hydrochloride (HH) – 15 % acetic acid – 0.03M Na-EDTA solution buffered to pH 4 with analytical grade NaOH for three hours in a shaker at room temperature. During initial stages of the study the Fe-Mn oxyhydroxide fractions were extracted using 0.5M HH or, alternatively, using an oxalate cocktail (Tovar-Sanchez et al., 2003). This approach was also successful for extracting the seawater Nd fraction from sediments, but more dilute leaching solutions were found to be sufficiently efficient to dissolve the seawater fraction. For analyses of the radiogenic isotope composition of the remaining detrital fractions a second leach with the above HH-acetic acid-Na-EDTA leach was applied for 24 hours to ensure complete removal of residual Fe-Mn oxyhydroxide coatings. Additionally two rinses with deionised water were carried out prior to treatment of the samples with *aqua regia* to destroy organic matter. The remaining sediment was dissolved by pressure digestion in a concentrated $HF-HNO_3$ mixture. Separation and purification of Nd then followed standard procedures (Cohen et al., 1988). Total procedural blanks for Nd in the oxyhydroxide fraction were <30 pg and 315 pg for the detrital fraction with all blanks being smaller than 1% of the total amount of Nd present in each sample.

Measurements of the Nd isotope compositions were carried out with a Nu Plasma MC-ICPMS at ETH Zürich. To correct for instrumental mass bias $^{143}Nd/^{144}Nd$ was normalised to a $^{146}Nd/^{144}Nd$ of 0.7219. All Nd isotope results were normalised to a $^{143}Nd/^{144}Nd$ of 0.512115 for the JNdi-1 standard (Tanaka et al., 2000). The long-term reproducibility for repeated measurements of the JNdi-1 standard was 0.27 ε_{Nd} (2σ) over a period of 16 months (n=70).

Depending on the composition of marine sediments sequential leaching of a pure seawater-derived fraction can potentially prove difficult because of unwanted chemical extraction of other phases. In the case of reductive leaching of Fe-Mn oxyhydroxide coatings from marine sediments the detrital fraction might be partially attacked and alter the extracted Nd isotope signal of the coatings. In earlier studies $^{87}Sr/^{86}Sr$ was measured on the same extracted Fe-Mn oxyhydroxide fractions to ensure the pure seawater origin of trace metals in the leachate (e.g. Rutberg et al., 2000; Piotrowski et al., 2005). Since Sr behaves conservatively in seawater, Sr extracted from Fe-Mn oxyhydroxides in pelagic sediments should in theory always reproduce the present-day $^{87}Sr/^{86}Sr$ seawater isotope ratio of 0.70918. This test was successfully applied to sediments from the South Atlantic (Rutberg et al., 2000; Piotrowski et al., 2005) but did not give a seawater Sr isotope composition for sediments at various locations in the central North Atlantic (Piotrowski, 2004).

For many samples in our study randomly elevated $^{87}Sr/^{86}Sr$ ratios were found, and these did not correlate with changes in the Nd isotopic composition. An extensive mass balance test, which is presented in chapter 2 showed that using the Sr isotope composition of the leachate appears to be a too strict monitor for the pure seawater origin of other radiogenic isotope systems such as Nd, mainly due to the very high Sr concentrations in the detrital fraction compared with other metals like Nd (as well as Pb and Th). According to these results, the detrital contribution to the Nd extracted by reductive leaching from the sediments along the Blake Ridge was found to be clearly outweighed by the Nd isotope signal of the authigenic fraction, no matter how radiogenic the measured Sr isotope signal.

3.2.2 $^{230}Th_{xs}$ and sediment redistribution

For the ^{230}Th-excess analyses bulk sediments were first treated with dilute HNO_3 and dried down before the addition of a $^{233}U/^{229}Th$ spike. After pretreatment with *aqua regia* and complete digestion in a concentrated $HF-HNO_3$ mixture, Th and U were separated and purified following the method given by Luo et al. (1997). Total procedural blanks for ^{232}Th are <10 pg and <15 pg for ^{238}U and are negligible. Procedural blanks for ^{230}Th and ^{234}U were below the detection limit of the secondary electron multipliers (SEMs) on the Nu Plasma MC-ICPMS.

Due to the cup configuration of the MC-ICPMS used for the $^{230}Th_{xs}$ analyses, individual measurements were split in two blocks to enable the simultaneous measurement of the isotopes ^{229}Th, ^{230}Th, ^{233}U and ^{234}U on two SEMs (Table 3.1). Using this configuration, the minor ^{234}U and ^{230}Th ion beams were monitored using IC0 and IC2 in the first analysis sequence, simultaneously with ^{235}U and ^{238}U on Faraday detectors, while ^{233}U and ^{229}Th were monitored subsequently, using the same multipliers, in a second analysis sequence (Table 3.1). The relative gain between the Faraday collectors and each SEM was calibrated against the CRM 145 U standard solution (formerly NIST SRM 960) doped with a ^{233}U/^{229}Th spike. Instrumental mass bias of the Th isotope measurements was corrected for externally by normalising to ^{238}U/^{235}U of 137.88 (Steiger and Jager, 1977). Linearly interpolated half mass zeros were subtracted from small ion beams to account for peak tailing from the large beams of the major isotopes ^{238}U and ^{232}Th.

Collector array for the Nu Plasma MC-ICPMS (Nu1) at ETH in Zürich.

	F(+2)	F(+1)	F(Ax)	F(-1)	F(-2)	**IC0(-3)**	F(-4)	IC1(-5)	F(-6)	**IC2(-7)**
Zero 1	237.5	236.5		234.5	233.5	**232.5**	231.5			**228.5**
Zero 2	238.5	237.5		235.5	234.5	**233.5**	232.5			**229.5**
Zero 3	239.5	238.5		236.5	235.5	**234.5**	233.5			**230.5**
Step 1		^{238}U			^{235}U	**^{234}U**				**^{230}Th**
Step 2	^{238}U			^{235}U		**^{233}U**	^{232}Th			**^{229}Th**

Table 3.1. Collector array for the Nu Plasma MC-ICPMS used for the U/Th analyses in Zürich. Simultaneous measurements of the different U and Th isotopes were carried out on Faraday collectors (F), and two secondary ion multipliers (IC0 and IC2). Numbers in brackets refer to the number of mass units (amu) away from the Axial Faraday collector. Three zero cycles were run for a better characterisation of the scattered ion background for ^{234,233}U and ^{230,229}Th.

The flux of unsupported ^{230}Th was calculated as outlined in Francois et al. (2004). Subsequently, following Suman and Bacon (Suman and Bacon, 1989), a focusing factor Ψ was determined, which is the ratio between the measured flux of ^{230}Th$_{xs}$ into the sediment and the expected ^{230}Th production rate in the overlying water column (i.e., 2.63 dpm / m^3 * kyr):

$$\psi = \frac{F(^{230}Th_{xs,initial})}{P_Z}$$

Where P_Z refers to the ^{230}Th production rate from the decay of authigenic ^{234}U per water depth z. If no syndepositional sediment redistribution occurred then $\Psi = 1$. If $\Psi > 1$ this indicates additional lateral transport and deposition of sediment (focusing) and if sediment is lost through bottom currents (winnowing) this results in $\Psi < 1$. Ψ $\gg 1$ are commonly found in drift deposits (Suman and Bacon, 1989; Thomson et al., 1993a). Focusing factors can only be determined between independently dated age tie points (Francois et al., 2004). The inherent potential error in the ^{230}Th-normalised flux method is on the order of 30 % at the highly dynamic location of the Blake Ridge, where advective transport of ^{230}Th must have occurred to a large extent. Individual results of the ^{230}Th$_{xs}$ measurements are displayed in Table 3.2.

The chronologies of the cores are based on published oxygen isotope stratigraphies displayed in Figures 3.4 and 3.5 (Keigwin, 2004). Published conventional ^{14}C ages (Keigwin, 2004) were transformed into calendar years using the marine radiocarbon age calibration Marine04 of Hughen et al. (2004) assuming $\Delta R = 0$.

3.3   Results

3.3.1   Flow dynamics, sedimentation rates and focusing

Today the DWBC is strongest along the lower Blake Outer Ridge between 3500 m and 4200 m water depth (Haskell and Johnson, 1993; Stahr and Sanford, 1999). Information on past modes of circulation at Blake Ridge is scarce. Mean grain size analyses on various sediment cores carried out earlier (Haskell, 1991) point to increased vigour of flow on the upper Blake Ridge during the deglaciation prior to 12 ka BP coinciding with reduced flow in the deeper parts. This pattern inverted after the Younger Dryas (Haskell, 1991). In cores below 2950 m water depth maximum mean grain sizes were recorded at ca. 10 ka BP, indicating maximum vigour of flow along the deeper Blake Ridge during the early Holocene (Haskell, 1991).

Sedimentation rates along the Blake Ridge varied systematically in response to the prevailing circulation patterns. Sedimentation rates at the site of 51GGC (1790 m) average 25.2 cm/kyr over the past 9000 yrs ((Keigwin, 2004); see Table 3.2), whereas sediments in the deeper parts of the Blake Ridge accumulated at ~ 4 cm/kyr during the Holocene (Luo et al., 2001). During the LGM the inverse situation is observed with sedimentation rates of 6.3 cm/kyr at shallow site 51GGC (Keigwin, 2004) and as high as 50 cm/kyr at the deeper locations (Luo et al., 2001).

$^{230}Th_{xs}$ results of core 51GGC, 1790 m

Depth in core (cm)	Calendar age (ka BP)	$^{238}U/^{232}Th$	$A(^{230}Th)$ (dpm/g)	$A(^{232}Th)$ (dpm/g)	$A(^{230}Th_{xs})_{meas}$ (dpm/g)	$A(^{230}Th_{xs})_{decay-corr}$ (dpm/g)	sedimentation rates (cm/kyr)	flux/kyr (dpm/g)*(cm/kyr)	
160	6.36	0.93390 ± 54	3.263	1.256	2.509	2.660	25.2	66.9	14.2
220	8.75	0.68364 ± 24	3.923	1.607	2.958	3.205	25.2	80.6	17.1
316	13.0	0.75168 ± 210	2.119	0.962	1.542	1.737	22.3	38.7	8.2
350	15.0	0.45632 ± 26	2.775	1.959	1.600	1.836	15.1	27.7	5.9
370	17.2	0.58286 ± 41	2.823	1.835	1.722	2.015	9.5	19.1	4.1
390	22.1	0.58770 ± 220	2.173	1.257	1.419	1.738	6.3	11.00	2.3
405	24.5	0.39025 ± 38	2.590	2.115	1.321	1.653	6.3	10.46	2.2
422	27.2	0.57571 ± 29	2.315	1.648	1.326	1.701	6.3	10.77	2.3

Table 3.2. Individual results for the $^{230}Th_{xs}$ analyses of core 51GGC presented in Figures 3.2 and 3.5. Associated errors represent the 2σ internal analytical error, quoted to two significant figures. All isotope ratios were corrected offline for the contribution from the trace amounts of natural ^{238}U, ^{235}U, ^{234}U, ^{232}Th and ^{230}Th present in the artificial spike tracer. Linearly interpolated half mass zeros were subtracted from all ion beams to account for the peak tailing contributions beneath the minor ion beams from the large beams of the major isotopes ^{238}U and ^{232}Th. Abundance sensitivity was <5 ppm amu^{-1} during the measurements (cf. Andersen et al. (2004)). Monitored $^{232}ThH^{+}$ formation rate was below 5 ppm of the ^{232}Th ion beam and subtracted from the ^{233}U ion beam. Sedimentation rates were calculated from Keigwin (2004). Secular equilibrium and CRM 145 U standard solution $^{234}U/^{238}U$ are taken from Cheng et al. (2000), $\lambda^{232}Th$ is taken from Holden (1990). The activities of the measured $^{230}Th_{xs}$ were decay-corrected to the time of deposition of the sediment using $\lambda^{230}Th$ of Cheng et al. (2000).

This variability in sedimentation rates in different parts of the ridge reflects the nature of this drift deposit with changing loci of sediment winnowing and focusing along the Blake ridge. The $^{230}Th_{xs}$ results indicate that during the LGM the shallow core 51GGC experienced only little sediment focusing (Ψ = 2.3; sedimentation rate: 6.3 cm/kyr), which increased during the deglaciation (Ψ = 4.1 to 8.2; sed. rate: 9.5 to 22.3 cm/kyr) and reached maximum values during the Holocene (average Ψ = 15.7; sed. rate averaging to 25.2 cm/kyr) (Fig. 3.2a, Table 3.2). Core CH88-11P in the lower segment of the Blake Outer Ridge, which was not subject of this study (Fig. 3.1) reveals the opposite pattern with most prominent sediment focusing during the LGM (Ψ of ca. 7.8; sed. rate: 27.3 cm/kyr) (Fig. 3.2b), decreasing focusing during the deglaciation (Ψ of ca. 4.7; sed. rate: 16.6 cm/kyr) and only minor focusing during the Holocene (Ψ of ca. 1.9; sed. rate: 3.7 cm/kyr) (calculated from data in Luo et al, 2001). The overall spatial variability of sediment focusing was most likely a direct function of the depth of the DWBC (Haskell, 1991; Haskell and Johnson, 1993), indicating that both sedimentation rates and the degree of sediment focusing were lowest within the main flow axis of the DWBC.

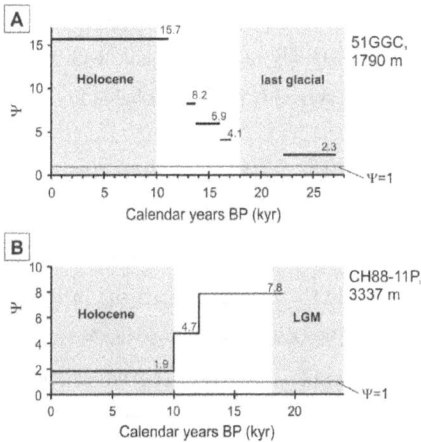

Figure 3.2. Focusing factor Ψ for (a) core KNR140, Site 51GGC analysed for $^{230}Th_{xs}$ activities in this study and (b) calculated for core CH88-11P analysed by Luo et al. (2001). The discontinuous records of the temporal evolution of the focusing factor arise from the fact that only average Ψ can be determined between absolute age tie points (cf. Francois et al., 2004). The shallow site experienced virtually no sediment focusing during the LGM and extreme focusing today, whereas the deep core reveals the opposite pattern. See text for discussion.

69

3.3.2   Neodymium isotope results

In Figure 3.3 the seawater Nd isotope compositions extracted from the Fe-Mn oxyhydroxide coatings are shown for the present-day situation (Fig. 3.3a) and the LGM (Fig. 3.3b). Direct seawater data obtained by Piepgras and Wasserburg (1987) from nearby sites southeast of the Blake Ridge are displayed for comparison. In the Holocene the Nd isotope composition of coatings measured from sites below ca 3200 m are identical to the seawater data. This segment of the Blake Outer Ridge represents the zone of highest along slope velocity of the DWBC today (Stahr and Sanford, 1999). The lowest ε_{Nd} is observed at 4200 m (ε_{Nd} of -13.5), coinciding with the highest DWBC speed observed by Haskell and Johnson (Haskell and Johnson, 1993). Below 4200 m the Nd isotope compositions become slightly more radiogenic again (ε_{Nd} = -12.5 in 4712 m water depth), most likely reflecting the influence of AABW at the deepest sites at the Blake Ridge (Stahr and Sanford, 1999). Above ca. 3200 m depth, the Nd isotope compositions define a systematic trend towards more radiogenic ε_{Nd} reaching highest values at 900 m water depth ($\varepsilon_{Nd} = -8.7$). During the LGM the unradiogenic Nd isotope signal characteristic of the present day LNADW Nd isotope composition was absent. Below 2200 m water depth the Nd isotope compositions define a straight line with a very stable ε_{Nd} of -10.5 ± 0.3. Above 2200 m water depth the LGM Nd isotopes were slightly more radiogenic with highest values of -9.5 in 1300 m water depth and are essentially indistinguishable from the Holocene data.

Downcore Nd isotope results for two sites from the lower part of the Blake Ridge are illustrated in Figure 3.4 plotted against depth in core, spanning the interval from the LGM to the present. In core 31GGC the pre-Holocene Nd isotope composition was essentially constant ($\varepsilon_{Nd} = -10.5 \pm 0.4$, Fig. 3.4a). A significant switch to ε_{Nd} values of -13, typical for present-day NADW occurred after the Younger Dryas followed by a slight increase to an ε_{Nd} value of -12.2 today. The Nd isotope evolution between the LGM and the Younger Dryas (12.9-11.5 ka BP) in deepest core 12JPC (Fig. 3.4b) appears somewhat more variable but the pattern is very similar to core 31GGC. The major hydrographic switch to typical LNADW-like Nd isotope compositions ($\varepsilon_{Nd} \sim -$ 13.5) was not initiated until after the Younger Dryas and there was also a slight increase during the Holocene towards the present day ε_{Nd} value of -13.0. None of the sampled depths in the two cores recorded LNADW-like ε_{Nd} until after the Younger

Dryas and in both cores the lowest ε_{Nd} was recorded shortly after the Younger Dryas.

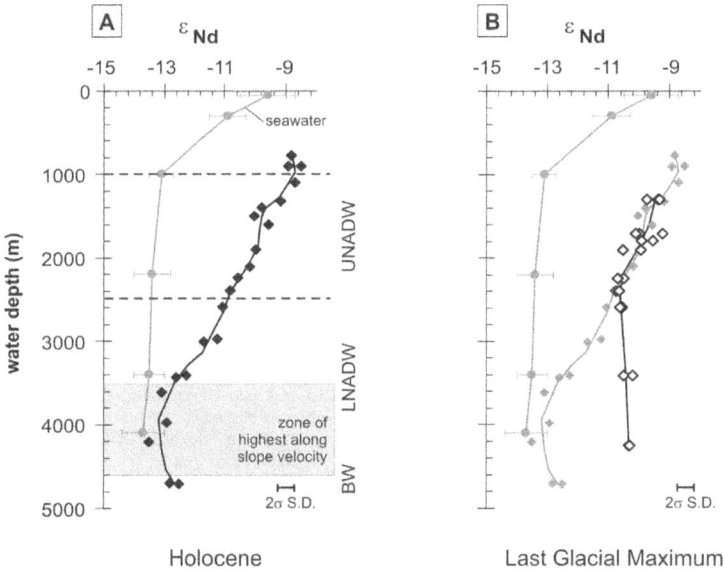

Figure 3.3. (a) Core-top Nd isotope seawater signal as recorded in Fe-Mn oxyhydroxide coatings in sediments along the Blake Ridge. Schematic present-day water mass structure is taken from (Stahr and Sanford, 1999) (UNADW: Upper North Atlantic Deep Water; LNADW: Lower North Atlantic Deep Water; BW: Bottom Water, i.e. mixture between NADW and Antarctic Bottom Water). Grey: Seawater measurements published by Piepgras and Wasserburg (1987) from station OCE63, located to the SSE of the Blake Ridge (see Fig. 3.1). (b) Neodymium isotope seawater signal during the LGM, measured on cores from nine different depths (open diamonds). Note that the prominent unradiogenic ε_{Nd} signature recorded in the core-top sediments below 2000 m is absent. The GNAIW ventilated depths between circa 1500 and 2000 m water depth. Individual Nd isotope data are presented in Table 3.3.

The Nd isotope composition of the detrital fraction in the sediments was measured at

three depths of core 12JPC (Fig. 4.4b). The detrital ε_{Nd} values have always been less

radiogenic than the oxyhydroxide coatings and range from −12.2 in the LGM to −15.0

at 11.2 ka BP.

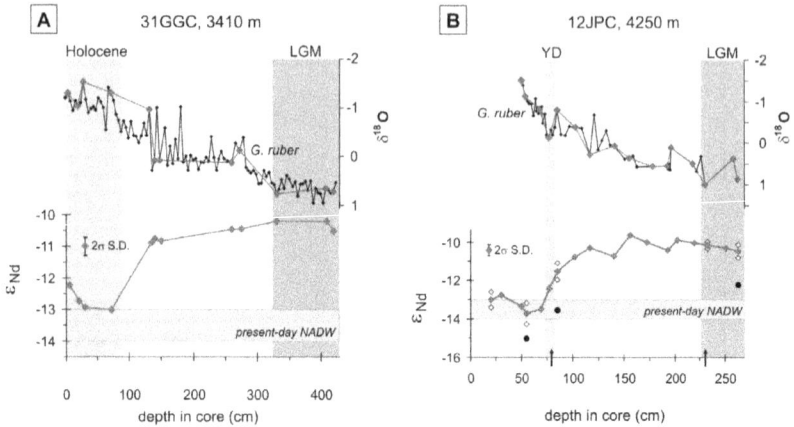

Figure 3.4. Downcore seawater Nd isotope results from this study and $\delta^{18}O$ stratigraphy from Keigwin (2004) (sampled depths in $\delta^{18}O$ plot are marked by red dots) for (a) core 31GGC in 3410 m and (b) 12JPC in 4250 m water depth along the deeper Blake Ridge. Blue arrows indicate absolute age tie points published in Keigwin (2004). Average results are plotted for depths in core for which duplicate analyses were carried out, however individual results are also shown (yellow filled diamonds). Dark grey circles represent Nd isotope compositions of detrital fractions. The modern NADW ε_{Nd}, as measured by Piepgras and Wasserburg (1987), is shown in grey box. Both cores did not record typical interglacial NADW signatures before the Holocene. Individual Nd isotope data are presented in Tables 3.3 to 3.5.

Shallow site (51GGC) from 1790 m water depth recorded a completely different Nd isotope evolution (Fig. 3.5). This location was chosen for downcore analyses in order to resolve the Nd isotope composition of GNAIW, which ventilated this depth during the LGM (Boyle and Keigwin, 1987; Sarnthein et al., 1994; Keigwin, 2004; Curry and Oppo, 2005). During the LGM only minor sediment focussing occurred ($\Psi = 2.3$) and the authigenic Fe-Mn oxyhydroxide signal ranged from an ε_{Nd} of –9.5 to –10.1, which is 3.5 to 4 ε_{Nd} units higher than interglacial LNADW at the deeper Blake Ridge. The sedimentary regime at this site changed at the end of the LGM from the deposition of coarse sand to silty sedimentation, and the time interval between 21 and 17.6 kyr BP is not preserved. Silt directly overlying the LGM section was deposited at 17.2 ka BP, the authigenic Fe-Mn oxyhydroxide fraction being less radiogenic than that in the LGM section. Compared with the LGM section the focusing factor in this early deglacial section increases significantly, contemporaneously with a drop in Nd isotope composition to an ε_{Nd} of –10.9. The Nd isotope signal remained unchanged until the Younger Dryas. During the following 2.5 kyr (i.e., the YD and the transition to the Holocene) the largest Nd-isotopic change occurred at this site with a shift of 1.7

ϵ_{Nd} towards more radiogenic compositions, coinciding with a pronounced enhancement of sediment focusing. Subsequently, Nd isotope compositions became slightly less radiogenic again until a present-day ϵ_{Nd} of -10.3 was reached.

At this site the ϵ_{Nd} of the detrital fractions is more radiogenic than at the deeper site and varies between -12.4 during the deglaciation and -10.8 during the Holocene (Fig. 3.5, 3.6).

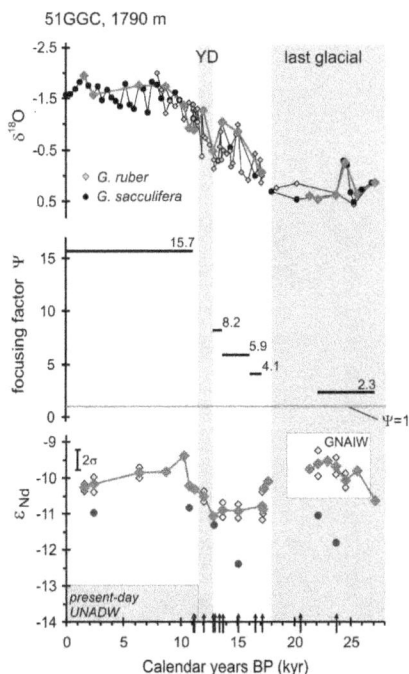

Figure 3.5. Upper panel: $\delta^{18}O$ stratigraphy from Keigwin (2004) (sampled depths in the $\delta^{18}O$ plot are marked by red dots). Middle panel: $^{230}Th_{xs}$ results as shown in Figure 3.2. The lower panel illustrates the Nd isotope evolution at site 51GGC, as recorded by Fe-Mn oxyhydroxide coatings. Average results are plotted for depths in core for which duplicate analyses were carried out but individual results are also shown (yellow filled diamonds). Neodymium isotope compositions of the detrital fraction is always less radiogenic than the Fe-Mn oxyhydroxide fraction in this core (dark grey filled circles). Individual Nd isotope data are presented in Tables 3.3 to 3.5. The modern NADW ϵ_{Nd}, as measured by (Piepgras and Wasserburg, 1987), is shown in grey box. Blue arrows indicate absolute age tie points published in Keigwin (2004).

In Figure 3.6 the seawater Nd isotopic evolution of the intermediate core 51GGC in 1790 m water depth is directly compared with that of the deepest core 12JPC in 4250 m water depth. The similarity between core 12 JPC and 51GGC before the onset of the Younger Dryas is apparent, as well as the switch to the presence of interglacial mode LNADW in core 12JPC after the Younger Dryas (cf. Fig. 3.4b). The intermediate core does not show UNADW-typical ε_{Nd} at any time during MIS 2 and 1. The isotopic compositions of detrital fractions in core 12JPC and 51GGC are also shown in Figure 3.6. Similar to the authigenic fraction, the detrital Nd isotope compositions did not vary significantly prior to the Y ounger Dryas but diverged afterwards.

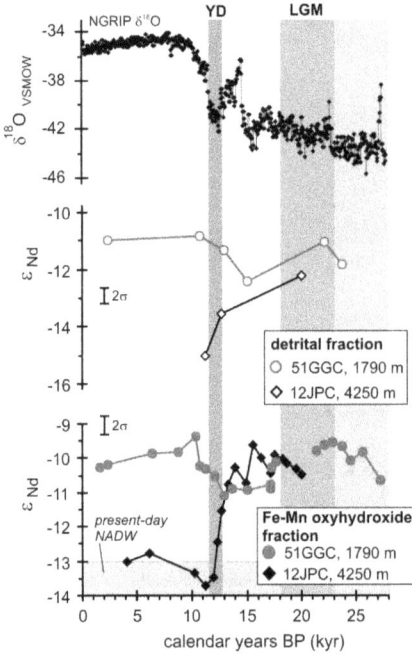

Figure 3.6.   Combined Nd isotope evolution of core 51GGC (1790 m) and 12JPC (4250 m), plotted against age. Also shown is the Northern Greenland $\delta^{18}O$ NGRIP record of Andersen et al. (2004) (upper panel), as well as the Nd isotope trends seen for the detrital fraction in the sediments (middle panel). The modern NADW ε_{Nd}, as measured by Piepgras and Wasserburg (1987), is shown in grey box.

Table 3.3a.   *Nd isotope results, Fe-Mn oxyhydroxide fractions, core-top samples*

Core	Water depth (m)	$^{143}Nd/^{144}Nd$	ε_{Nd}
GGC-68	775	0.512186 ± 6	-8.8
JPC-63	900	0.512186 ± 6	-8.8
JPC-63	duplicate	0.512190 ± 19	-8.7
GGC-62	906	0.512181 ± 9	-8.9
GGC-60	1100	0.512192 ± 8	-8.7
GGC-57	1323	0.512168 ± 6	-9.2
GGC-56	1400	0.512137 ± 19	-9.8
GGC-54	1495	0.512123 ± 5	-10.0
GGC-53	1605	0.512148 ± 6	-9.6
GGC-50	1903	0.512127 ± 7	-10.0
GGC-64	2101	0.512116 ± 5	-10.2
PG-1	2243	0.512097 ± 10	-10.6
PG-2	2394	0.512084 ± 7	-10.8
GGC-43	2590	0.512071 ± 6	-11.1
GGC-39	2975	0.512062 ± 6	-11.2
JPC-36	3007	0.512040 ± 15	-11.7
GGC-30	3433	0.511992 ± 7	-12.6
GGC-32	3615	0.511966 ± 7	-13.1
GGC-29	3978	0.511976 ± 7	-12.9
GGC-28	4211	0.511945 ± 6	-13.5
GGC-21	4705	0.511980 ± 8	-12.8
JPC-22	4712	0.511998 ± 21	-12.5

Table 3.3b.   *Nd isotope results, Fe-Mn oxyhydroxide fractions, LGM samples*

Core	Depth in core (cm)	Water depth (m)	$^{143}Nd/^{144}Nd$	ε_{Nd}
IODP 172 1054A 001 04W -				
42-44	492 cm	1302	0.512157 ± 6	-9.4
52-54	502 cm	1302	0.512160 ± 10	-9.3
62-64	512 cm	1302	0.512139 ± 8	-9.7
KNR140/2 -			±	
52GGC	395 cm	1710	0.512126 ± 5	-10.0
52GGC	405 cm	1710	0.512120 ± 7	-10.1
52GGC	385 cm	1710	0.512165 ± 8	-9.2
50GGC	260 cm	1903	0.512129 ± 5	-9.9
50GGC	270 cm	1903	0.512098 ± 7	-10.5
01JPC	167 cm	2243	0.512100 ± 6	-10.5
01JPC	183 cm	2243	0.512090 ± 6	-10.7
02JPC	21 cm	2394	0.512087 ± 4	-10.7
02JPC	28 cm	2394	0.512093 ± 5	-10.6
43GGC	182 cm	2590	0.512098 ± 5	-10.5
43GGC	187 cm	2590	0.512094 ± 6	-10.6

Table 3.3.   Individual Nd isotope results for the Fe-Mn oxyhydroxide fraction of core-top and LGM sediments presented in Figure 3.3 including the 2σ internal errors.

Table 3.4.   *Nd isotope results, Fe-Mn oxyhydroxide fractions*

Depth in core (cm)	Calendar Age (ka)	$^{143}Nd/^{144}Nd$		ε_{Nd}
51GGC, 1790 m				
40	1.59	0.512107	± 5	-10.4
40	duplicate	0.512117	± 5	-10.2
60	2.39	0.512106	± 4	-10.4
160	6.36	0.512126	± 6	-10.0
160	duplicate	0.512140	± 6	-9.7
220	8.75	0.512135	± 6	-9.8
260	10.34	0.512157	± 8	-9.4
290	11.20	0.512109	± 6	-10.3
300	12.06	0.512092	± 5	-10.7
300	duplicate	0.512107	± 8	-10.4
310	12.89	0.512070	± 6	-11.1
330	13.72	0.512069	± 5	-11.1
330	duplicate	0.512091	± 6	-10.7
350	15.0	0.512068	± 5	-11.1
370	17.2	0.512065	± 5	-11.2
370	duplicate	0.512106	± 5	-10.4
371	17.2	0.512074	± 6	-11.0
371	duplicate	0.512085	± 5	-10.8
373	17.3	0.512110	± 6	-10.3
378	17.7	0.512121	± 8	-10.1
385	21.4	0.512137	± 7	-9.8
390	22.1	0.512128	± 8	-10.0
395	22.9	0.512149	± 10	-9.5
400	23.7	0.512130	± 7	-9.9
405	24.5	0.512111	± 8	-10.3
405	duplicate	0.512132	± 6	-9.9
412	25.6	0.512135	± 7	-9.8
422	27.2	0.512093	± 7	-10.6
31GGC, 3410 m				
core-top		0.512011	± 25	-12.2
20		0.511985	± 5	-12.7
30		0.511974	± 5	-12.9
72		0.511971	± 5	-13.0
135		0.512080	± 8	-10.9
140		0.512087	± 5	-10.8
150		0.512083	± 6	-10.8
260		0.512101	± 5	-10.5
275		0.512103	± 6	-10.4
330		0.512115	± 5	-10.2
410		0.512115	± 5	-10.2
420		0.512099	± 5	-10.5

Table 3.4.   Individual Nd isotope results for the Fe-Mn oxyhydroxide fraction for sediments of core 51GGC, 31GGC and 12JPC presented in Figures 3.4, 3.5 and 3.6 including the 2 σ internal errors.

Table 3.4. Continued.

Depth in core (cm)		$^{143}Nd/^{144}Nd$		ε_{Nd}
12JPC, 4250 m				
20		0.511951	± 6	-13.4
20	duplicate	0.511992	± 4	-12.6
30		0.511984	± 6	-12.8
50		0.511954	± 6	-13.3
55		0.511907	± 8	-14.3
69		0.511947	± 6	-13.5
77	YD	0.512001	± 6	-12.4
85		0.512025	± 6	-12.0
102		0.512087	± 6	-10.8
117		0.512111	± 5	-10.3
140		0.512088	± 5	-10.7
156		0.512145	± 5	-9.6
172		0.512126	± 6	-10.0
193		0.512104	± 5	-10.4
203		0.512130	± 6	-9.9
220		0.512124	± 7	-10.0
233	LGM	0.512108	± 6	-10.3
233	duplicate	0.512126	± 5	-10.0
251		0.512110	± 4	-10.3
263		0.512083	± 7	-10.8

Table 3.5. Samples for which both the coatings and the detrital fraction were analysed

Core	Depth in core (cm)	sample type	$^{143}Nd/^{144}Nd$		ε_{Nd}
51GGC	60	coating	0.512127	± 12	-10.0
		detritus	0.512076	± 10	-11.0
51GGC	270	coating	0.512114	± 10	-10.2
		detritus	0.512083	± 9	-10.8
51GGC	316	coating	0.512109	± 10	-10.3
		detritus	0.512057	± 10	-11.3
51GGC	350	coating	0.512088	± 11	-10.7
		detritus	0.512003	± 8	-12.4
51GGC	390	coating	0.512164	± 9	-9.3
		detritus	0.512071	± 10	-11.1
51GGC	400	coating	0.512154	± 9	-9.4
		detritus	0.512033	± 9	-11.8
12JPC	55	coating	0.511963	± 7	-13.2
		detritus	0.511868	± 8	-15.0
12JPC	85	coating	0.512069	± 9	-11.1
		detritus	0.511943	± 9	-13.6
12JPC	263	coating	0.512119	± 9	-10.1
		detritus	0.512012	± 7	-12.2

Table 3.5. Individual Nd isotope results for the Fe-Mn oxyhydroxide and the detrital fractions for sediments of core 51GGC and 12JPC presented in Figures 3.4b, 3.5 and 3.6 including the 2σ internal errors. All but two (i.e., 51GGC-270; 51GGC-316) Fe-Mn oxyhydroxide coating results shown here represent additional duplicates to samples plotted in Table 3.4.

3.4   Discussion

3.4.1   Present day Nd isotope distribution at the Blake Ridge

Below 3200 m water depth the Nd isotope signal of Fe-Mn oxyhydroxide coatings match that of present-day seawater confirming LNADW presence along the deeper Blake Ridge today. Above the zone of highest along slope velocity of the DWBC the authigenic Nd isotope compositions obtained from the Fe-Mn oxyhydroxide coatings become systematically more radiogenic with decreasing water depth and deviate from direct seawater measurements from nearby water sampling sites (station OCE63; Piepgras and Wasserburg, 1987) by up to 4 ε_{Nd} units in 1000 m water depth. Detrital contamination of the extracted seawater Nd isotope signal can be ruled out as a cause for this offset. Due to mass balance constraints unrealistically radiogenic Nd isotope compositions of the detrital fraction - even more radiogenic than present-day MORB - would be required to alter the Nd isotope signal to such an extreme extent (chapter 2). The Nd isotope signal of the detrital fraction in all studied cores along the Blake Ridge, however, is always less radiogenic than that of the Fe-Mn oxyhydroxide phase (Figs. 3.4-3.6).

Given that no water column Nd isotope data are available from directly above the Blake Ridge, it is possible, though unlikely, that the water masses and corresponding isotope compositions along the Blake ridge are significantly different from the open ocean profile studied at the OCE63 stations. This needs to be entertained, however, as one possible way to interpret the data and could only be unambiguously disproved by future water column measurements in today's water column at the sediment – bottom water interface.

The Nd isotope compositions presented in Figures 3.3 and 3.5 are in conflict with direct seawater measurements published earlier (Piepgras and Wasserburg, 1987; van de Flierdt, 2006) but in very good agreement with dissolved and particulate Nd isotope compositions of nearby eastern North America (Goldstein and Jacobsen, 1987). Only North American shelf water has similar ε_{Nd} to seawater compositions recorded in Fe-Mn oxyhydroxide coatings at intermediate depths along the Blake Ridge. Therefore the question arises how such a shallow water signal can be exported to abyssal depths.

If the present-day water column Nd isotope signatures along the Blake Ridge are indeed truly represented by the OCE63 values (Piepgras and Wasserburg, 1987) in Figure 3.3a, this would support the suggestion of particle-seawater Nd isotope exchange through the boundary exchange mechanism at the continent-ocean interface (Lacan and Jeandel, 2005b). The only way to modify the characteristic Nd isotope composition of NADW along the flow path of the DWBC along eastern North America would be via sediment-seawater interaction along this continental rise. The average bulk crustal ε_{Nd} along eastern North America is –9 (Goldstein and Jacobsen, 1987) and thus such an alteration of the original NADW appears possible. A closer look at Table 3.4, Figures 3.4b and 3.5 reveals, however, that ε_{Nd} of the detrital fraction in the sampled depths varies between –10.8 and –15.0. The Nd isotope composition at any sampled depth is always less radiogenic than the Fe-Mn oxyhydroxide fraction in the respective depths. Therefore, it is unlikely that boundary exchange has been the driving mechanism of altering the seawater values generating the observed Nd isotope pattern recorded in Fe-Mn oxyhydroxide coatings.

3.4.2   A plausible mechanism to export a surface water signal

The upper continental slope above the Blake Ridge is the location of a major water mass boundary (Figs. 3.1, 3.7). To the west of the Blake Ridge on the Blake Plateau the vigorously flowing Gulf stream currently transports approximately 30 Sv of water derived from the Florida current northwards (Schmitz and McCartney, 1993; Reverdin et al., 2003). Interaction of Gulf stream water with near shore water masses is complex and variable, and leads to turbulence and re-suspension of sediments deposited on the shelf and upper continental slope (Eittreim et al., 1969; Eittreim et al., 1976; McCave, 1986; Biscaye et al., 1988). As the Gulf Stream passes the North American shelf edge towards the open Atlantic, re-suspended sediment from the shelf and upper continental slope is swept off the shelf towards the open Atlantic in a prominent intermediate depth nepheloid layer in the western North Atlantic (Eittreim et al., 1969; McCave, 1986; Biscaye et al., 1988). Significant amounts of re-distributed sediment on the Blake Ridge must be derived from shallow shelf sites (Hunt et al., 1977; Biscaye et al., 1994). Because the major along slope flow axis of the DWBC is centred between 3500 m and 4200 m water depth, the bulk of the exported shelf and continental slope sediments were most likely deposited on the upper continental rise above the core of the DWBC, which limits the deposition of sediment.

During temporary storage on the shelf and prior to re-suspension, the sediments acquired a first generation of Fe-Mn oxyhydroxide coatings incorporating the Nd isotope signature of the local surface waters. When these sediments are exported downslope to the Blake Ridge and deposited at high sedimentation rates, this original surface water mass Nd isotope signal was also re-deposited and hence contributes to varying degrees to the extracted in situ deep water signals. The deeper the sampling sites along the Blake Ridge, the less prominent the shallow water bias in the Fe-Mn oxyhydroxide fraction. Original NADW Nd isotope compositions are attained in the core of the DWBC below 3200 m, where fallout from the intermediate nepheloid layer cannot be deposited due to the high current speed.

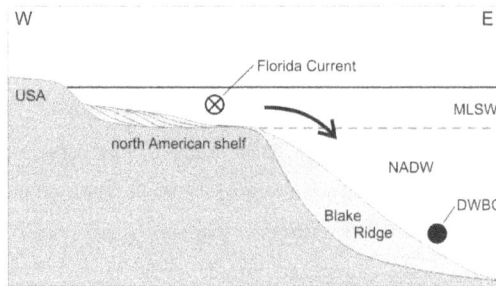

Figure 3.7.   Schematic simplified cross section through the NE American continental shelf and continental rise towards the open North Atlantic. Sediments with pre-acquired Fe-Mn oxyhydroxide coating are prone to re-suspension on the North American shelf, as indicated by the trajectory of the vigorously flowing Florida Current, and sediment is exported into depths where NADW prevails. The filled black dot marks the approximate position of the major flow axis of the DWBC today. MLSW: Mid-latitudinal warm surface water. See text for discussion.

3.4.3   Deep water Nd isotope evolution

The highly unradiogenic Nd isotope signature below 3200 m depth obtained from the Holocene Fe-Mn oxyhydroxide coatings in cores 31GGC and 12JPC (Fig. 3.4) was absent during the LGM (Fig. 3.3b, 3.4, 3.6). Seawater ε_{Nd} in the cores along the Blake Ridge obviously reflects the presence of a different water mass at the deeper Blake Ridge during the LGM. Various authors suggested a stronger presence of SSW at depths below 2000 to 2500 m water depth in the glacial North Atlantic (Boyle and Keigwin, 1987; Labeyrie et al., 1992; Marchitto et al., 2002; Keigwin, 2004; Curry and Oppo, 2005; Robinson et al., 2005), which is fully consistent with the Nd isotope data. The observed ε_{Nd} signature of -10.3 ± 0.5 suggests that glacial SSW, which was in the ε_{Nd} range of -6 to -7 in the Southern Atlantic (Jeandel, 1993; Abouchami et al., 1999; Rutberg et al., 2000; Piotrowski et al., 2004), was already modified significantly by mixing with waters of North Atlantic origin by the time it reached the deeper Blake Ridge, unless the AAIW ε_{Nd} during the LGM was less radiogenic than today.

Figures 3.4 and 3.6 show that the ε_{Nd} of the water mass present at 4250 m at Blake Ridge during the LGM remained unchanged throughout the deglaciation until the Younger Dryas. Core 31GGC from 3410 m water depth shows strikingly similar patterns but due to lower sampling resolution any short-term excursions, as indicated by radiocarbon data (Robinson et al., 2005), are not resolved in our record. What can be resolved, however, is that no major water mass change of a duration longer than a few hundred to a thousand years was recorded at deeper sites along the Blake Ridge until after the Younger Dryas. The authigenic Nd isotope record from site 12JPC in Figure 3.6 suggests that the change towards interglacial mode LNADW compositions occurred already during the Younger Dryas. This observation is most likely due to very low sedimentation rates since the latest Younger Dryas (<5cm/kyr), leading to slight offsets between depositional age and the time of Fe-Mn oxyhydroxide formation (i.e., lock-in depth of the authigenic Nd isotope signal, which is on the order of a few centimetres below the seafloor). We therefore propose that the actual change in Nd isotope composition of the water mass prevailing at 12JPC occurred after the Younger Dryas.

3.4.4   GNAIW Nd isotope composition

Because seawater ε_{Nd} at a particular location is a function of the flow path of the prevailing water mass and potential continental contributions in its source areas (Frank, 2002; Goldstein and Hemming, 2003), a major unknown in the reconstruction of Atlantic meridional overturning circulation during the LGM using Nd isotopes is the ε_{Nd} signature of the glacial equivalent of NADW, the GNAIW. During the last glacial site 51GGC was bathed in GNAIW (highlighted in Fig. 3.5). During at least 4 kyr the Nd isotope composition of the GNAIW recorded in Fe-Mn oxyhydroxide coatings remained relatively constant at an average ε_{Nd} of –9.8. This is approximately 3.5 to 4 ε_{Nd} different from the interglacial NADW. The grain size of the sediment at this depth in the core is dominated by sand indicating strong current velocities. Additionally, the $^{230}Th_{xs}$ data show that this site was not significantly affected by sediment focussing during the LGM (Fig. 3.5).

If we have indeed correctly identified the process that affected the Holocene and deglacial Nd isotope compositions extracted from Fe-Mn oxyhydroxide coatings in core 51GGC, then this mechanism was not operating during the LGM (i.e., virtually no sediment focussing). In this case, although similar to the Holocene compositions, the Nd isotope signature presented here for the intermediate water depth of 1790 m should reflect the true GNAIW Nd isotope signal (Fig. 3.5). The average GNAIW ($\varepsilon_{Nd} \approx -9.7$) along the Blake Ridge suggests that waters circulating through the Labrador Sea did not contribute to the water mass signal recorded at the Blake Ridge. Even minor amounts of seawater derived from the Labrador Sea should significantly lower the observed ε_{Nd} along intermediate depths at the Blake Ridge.

Whether Labrador Sea water contributed to the DWBC during the LGM is still under debate. Rasmussen et al. (2003) suggested that LNADW, which formed in the Norwegian-Greenland Sea reached the SE Labrador Sea between Heinrich events based on oxygen and carbon isotope data, as well as IRD data and faunal distributions. It has also been suggested that at present Labrador Sea Water is formed at two localities, the first being located in the Labrador basin and the second region being located in the southwest Irminger Sea (Pickart et al., 2003; Falina et al., 2007). Hence if Labrador deep water formation today occurs at various localities, then this hydrographic situation could possibly also apply to the LGM, thereby allowing waters from the Labrador Sea with unradiogenic ε_{Nd} to be mixed and incorporated into GNAIW. Conversely, transfer functions based on dinocysts and $\delta^{18}O$ in planktonic foraminifera (de Vernal and Hillaire-Marcel, 2000; Hillaire-Marcel et al., 2001; de Vernal et al., 2002) strongly argue for the development of a strong pycnocline in the Labrador Sea during the LGM, conditions highly unfavourable for vertical convection. The very radiogenic GNAIW ε_{Nd} results presented here suggest a complete shutdown of Labrador Sea Water formation during the LGM. Furthermore, if waters from the Norwegian-Greenland Sea remained constant in Nd isotope composition between the LGM and today then Greenland Sea Intermediate and Deep Waters ($-10.7 \leq \varepsilon_{Nd} \leq -10.3$) (Lacan and Jeandel, 2004a) are also not likely to contribute to GNAIW. In the light of recently emerging Nd isotope data (van de Flierdt, 2006; Foster et al., 2007) this suggested shutdown must have been of short duration, possibly being restricted to the LGM or the entire MIS2, which needs to be confirmed in future studies.

3.4.5 Reduced export of NADW to the Southern Ocean (Cape Basin) during the LGM?

A major conclusion of the high-resolution study of sediment core RC11-83 by Piotrowski et al. (2004, 2005) was a major decrease of glacial and stadial NADW export to the Cape Basin based on a shift of three ε_{Nd} units to -6.4 in the LGM compared with -9.3 during the Holocene. According to these calculations the Pacific water component increased by approximately 50 % relative to the Atlantic component during times of lowest NADW export (Piotrowski et al., 2004). Should the results presented here for the Nd isotope composition indeed reflect the GNAIW ε_{Nd} values this estimate needs to be adjusted. Our data suggest that along the Blake Ridge the GNAIW during the LGM was 3.5 to 4 ε_{Nd} different from interglacial UNADW, which then explains a significant part of the observed shift in ε_{Nd} in the Cape Basin between the LGM and the Holocene without invoking changes in the export of NADW/GNAIW.

Based on Nd isotope compositions of seawater in the Drake Passage, Piepgras and Wasserburg (1982) estimated the Antarctic Circumpolar Deep Water (CDW) today to be composed of 50 to 70 % Atlantic water. By maintaining this mixing ratio the change in GNAIW composition reported here can account for more than half of the observed glacial-interglacial ε_{Nd} variation of CDW in the Cape Basin (1.5 to 2.5 ε_{Nd}), but not for the entire shift published earlier Piotrowski et al., 2004; 2005). A significant reduction of glacial NADW (GNAIW) export to the Southern Ocean is still required to generate the observed shift towards more Pacific-like glacial Nd isotope compositions of CDW in the Cape Basin, assuming that the Nd concentrations of the glacial water masses were the same as those observed today.

3.5   Conclusions

For core-top sediments along the Blake Ridge the original LNADW Nd isotope composition can be extracted reliably from Fe-Mn oxyhydroxide coatings in sediments from within the major flow axis of the DWBC. Above the major flow axis the extracted Nd isotope signal from Fe-Mn oxyhydroxide coatings of sediments appears to be increasingly offset towards a surface seawater Nd isotope signal from the North American shelf, caused by the re-distribution of dispersed shelf sediments downslope the Blake Ridge. Today a slight influence of Antarctic Bottom Water is observed in the Fe-Mn oxyhydroxide coating signals at the lowermost sites at 4700 m water depth along the Blake Ridge.

The Nd isotope composition of GNAIW measured along the Blake Ridge is about 3.5 to 4 ε_{Nd} units higher than interglacial LNADW. Although similar to Holocene compositions ($\varepsilon_{Nd, GNAIW}$ = -9.7 ± 0.4), the Fe-Mn oxyhydroxide coatings of the glacial section in 1790 m water depth were not biased by sediment focussing as demonstrated by $^{230}Th_{xs}$ data. Therefore the Nd isotope compositions measured for the LGM section are considered to reflect the true ambient seawater composition at this water depth. According to our data, the distinctly unradiogenic Nd isotope composition typical of present day NADW was thus not present during the LGM. As a consequence, it is not likely that water derived from the Labrador Sea contributed significantly to GNAIW during the LGM. The radiogenic ε_{Nd} of the GNAIW suggests that Nd originating from isotopically unmodified waters from either the Irminger- or the Iceland Basins in the subpolar North Atlantic fed the GNAIW along the upper Blake Ridge. The ε_{Nd} value of the Southern Source-influenced deep waters at the lower Blake Ridge was circa – 10.5 during the LGM.

The Nd isotopes in sediment cores from water depths below 3400 m did not record the presence of typical LNADW until after the Younger Dryas. Therefore the flow of the DWBC must have been restricted to the upper segment of the Blake Ridge, and modern circulation was not initiated until after the Younger Dryas. Comparison of the Blake Ridge data with Nd isotope reconstructions from the Cape Basin suggest that there was a significant reduction in the export of North Atlantic waters during the glacial, however, not as strong as previously estimated.

Acknowledgements

M. B. Andersen, E.-K. Potter and M. Wipf are acknowledged for providing assistance with the Th separation and measurements. Tina van de Flierdt, Brian Haley, Sidney Hemming and Gavin Foster are thanked for additional constructive suggestions.

Chapter 4

Seawater Pb isotopes in the western North Atlantic: Laurentide ice sheet decay, continental freshwater runoff diversions and the establishment of the Holocene weathering regime

Abstract

The isotopic composition of seawater-derived Pb has been extracted from marine drift deposits along the Blake Ridge in the western North Atlantic in order to resolve changes in the continental weathering flux of Pb to the ocean at sub-millennial resolution. Three cores were studied from shallow (1790 m), intermediate (3410 m) and deep (4250 m) water depths and yet unrecognised short-term changes in the Pb isotopic composition of seawater could be identified. The Pb isotope signal in the western North Atlantic switched from relatively unradiogenic isotopic compositions during the Last Glacial Maximum (LGM) and the subsequent deglaciation to highly radiogenic compositions after the Younger Dryas. The Pb isotope compositions have levelled out to slightly less radiogenic isotopic compositions today. Lead is thought to be released incongruently during weathering with the net effect that the Pb isotope composition of the weathering does not correspond to that of the source rocks, making it basically impossible to constrain the continental source area by means of the various Pb isotope ratios alone. The seawater Pb isotope patterns shown here reflect the slow retreat of the Laurentide ice sheet, which retained a large volume on the eastern Canadian Shield until after the Younger Dryas. While the ice sheet was still substantial, continental freshwater runoff was directed in the Gulf of Mexico. Only when the ice sheet was retreated far enough towards the north could freshwater be directly supplied into the western North Atlantic, giving rise to more radiogenic Pb isotope compositions through more direct freshwater inputs. The most radiogenic Pb isotope composition was recorded in the western North Atlantic at about 11.2 ka BP. Our records also suggest that the continental Pb input in the North Atlantic was significantly reduced during the LGM, and that long-term Pb isotope trends towards more radiogenic compositions recorded in ferromanganese crusts in fact reflect short-lived interglacial Pb isotope signals recorded during otherwise prevailing ice-house conditions, implying that the majority of continental radiogenic Pb was supplied to the North Atlantic during interglacials. The most dramatic Pb isotope excursion occurred over a period of less than 3 kyr and for this reason it could not be resolved in earlier studies using ferromanganese crusts. The consistent large amplitude Pb isotope variations recorded on sub-millennial timescales demonstrate that Pb isotope compositions recorded by Fe-Mn oxyhydroxide coatings of marine sediments are a powerful tool for high resolution reconstructions of past climate, continental runoff and weathering regimes.

4.1   Introduction

The Pleistocene (1.8 Ma to 10 ka BP) was an extreme climatic period in which the geomorphology of the mid and high latitudes in both hemispheres was strongly modulated by a continuous series of glacial advances and interglacial retreats of continental ice sheets involving efficient continental denudation during glacial erosion. Interstadials, representing short intervals (500 to 2,000 years) (Johnsen et al., 1992) of relatively mild climatic conditions within the glacials, are enigmatic features in this "ice-house climate", and although these brief warm episodes are well resolved in ice core records (Dansgaard et al., 1993; Petit et al., 1999) it is not clear to what extent the climate during these periods really differed from truly glacial conditions. Continental climate can be traced by means of weathering fluxes, because the effectiveness of chemical and physical weathering is closely coupled to mean annual temperatures. Other proxies have been employed to trace climate regulated continental weathering fluxes such as the Chemical Index of Alteration (CIA) (Nesbitt and Young, 1982; Nesbitt and Young, 1984), $^{87}Sr/^{86}Sr$ isotopic compositions in carbonate materials within the oceans (Raymo and Ruddiman, 1992), the combination of Os, Nd and Pb isotope systems in seawater records spanning the past 70 Myr (Burton, 2006), and U-series disequilibria induced by elemental fractionation during chemical weathering (Dosseto et al., 2006).

The dissolved Pb flux transferred from the continents to the oceans and the related changes in Pb isotopic composition of seawater is a particularly promising indicator of weathering, and earlier studies on ferromanganese crusts in the North Atlantic already ascribed significant variations in Pb isotope composition to changes in the weathering flux into the oceans during the Pleistocene (Reynolds et al., 1999). Unlike with Nd, the Pb isotopic composition of seawater does not simply reflect the bulk source composition of a continental source area, but is released incongruently during weathering on the continents (von Blanckenburg and Nagler, 2001; Harlavan and Erel, 2002). During the last glacial cycle the Laurentide ice sheet created extensive quantities of glacially eroded fresh rock substrate. Throughout the Last Glacial Maximum this moraine material could not be chemically weathered to any significant extent as the effectiveness of chemical and physical weathering is closely coupled to mean annual temperatures. Intensified chemical weathering initiated in the course of the deglaciation, and because of this feature dissolved Pb fluxes resolved at sub-millennial timescale are particularly useful. Rock substrate in form of moraine material is a mixture of more or less weathering-resistant mineral phases. Accessory U- and Th-rich mineral phase are particularly prone to chemical disintegration and earlier studies demonstrated a preferential release of highly radiogenic Pb during early chemical weathering (Harlavan et al., 1998; Harlavan and Erel, 2002). We can utilise the Pb specific weathering behaviour and trace continental runoff pathways into the ocean during the deglaciation by measuring the Pb isotope composition of paleo-seawater archives such as the authigenic Fe-Mn oxyhydroxide fraction of marine sediments.

Lead is very particle reactive with a residence time in seawater in the range of 50 years in the Atlantic (Henderson and Maier-Reimer, 2002) and below 200-400 years in the Pacific (Craig et al., 1973; Schaule and Patterson, 1981; Frank, 2002). This is significantly shorter than the rate of Atlantic and global overturning circulation. Because of its short residence time in seawater it has been argued that Pb is not advected over long distances between ocean basins and merely reflects a local input signal, which can, however, serve to reconstruct short distance water mass mixing (Abouchami and Goldstein, 1995; von Blanckenburg et al., 1996). So far the Pb isotope composition in ferromanganese crusts has been used to infer long-term changes in deep water circulation (Christensen et al., 1997; Abouchami et al., 1999; Frank et al., 1999a; Reynolds et al., 1999; Frank et al., 2002) and reorganisations of oceanic gateways (Burton et al., 1997; Frank et al., 1999b). Short-term changes of the past dissolved Pb isotope composition of seawater have up to now not been accessible due to missing suitable archives.

The dissolved marine Pb isotopic composition is dominated by dissolved and particulate riverine (Chow and Patterson, 1962; O'Nions et al., 1978), eolian (Jones et al., 2000) or deep sea hydrothermal (van de Flierdt et al., 2004c) inputs. Due to its effective scavenging from seawater and strikingly similar Pb isotope compositions of abyssal ferromanganese crust in the different ocean basins it has also been argued that the hydrogeneous Pb isotope signal the abyssal deep sea is dominated by continuous atmospheric volcanic aerosol input (Klemm et al., 2007). While past natural seawater Pb isotope trends can be determined the present-day seawater Pb signal is dominated by atmospheric anthropogenic Pb input (Schaule and Patterson, 1981; Alleman et al., 1999).

In order to overcome the restriction in temporal resolution of ferromanganese crust records to reconstruct past seawater composition, Foster and Vance (2006) recently published high resolution laser ablation Pb isotope results from ferromanganese crusts in the western North Atlantic. The glacial-interglacial Pb isotope evolution over the last 550 kyr was obtained at a temporal resolution of around 10 kyr. Despite the higher resolution achieved by these authors compared with previous work, potential millennial and sub-millennial Pb isotope variations are still not resolvable and the age control is limited. Considering the expected Pb isotope variations occurring during short-lived paleoceanographic and climatic perturbations such as the deposition of ice rafted detrital material during Heinrich events (Hemming, 2004), a new paleoceanographic seawater archive needs to be employed to achieve the resolution required to investigate causal and temporal relations between events. Marine sediments deposited at relatively high sedimentation rates contain early diagenetic authigenic Fe-Mn oxyhydroxides that incorporate the trace metal isotope composition of the overlying deep water mass. After chemical extraction through reductive leaching the Nd isotope composition of this archive was used to resolve sub-millennial changes in North Atlantic Deep Water (NADW) export into the Cape Basin of the South Atlantic (Piotrowski et al., 2005). Because seawater-derived Pb is also stored in these Fe-Mn oxyhydroxide coatings it can be extracted following the same procedures as for Nd. This study represents the first reconstruction of the seawater Pb isotope evolution on sub-millennial timescales, and is aimed at reconstructing paleoclimatic and weathering-related changes during the transition from the Last Glacial Maximum to the Holocene. Particular emphasis is put on sub-millennial changes in the seawater Pb isotope compositions over the course of the Younger Dryas (YD).

4.2   Material and Methods

Three cores retrieved during KNR140 cruise along the Blake Ridge in the western North Atlantic spanning the interval from the LGM to present-day were selected for Pb isotope analyses. Core 51GGC from 1780 m depth, 31GGC from 3410 m depth and 12JPC from 4250 m depth were sampled to monitor the seawater Pb isotope evolution in the shallow, intermediate and deep western North Atlantic, respectively (Fig. 4.1). Only bulk sediments were used because tests at early stages of the study indicated easily disturbed Pb isotope signals due to blank contributions if sediments were sieved prior to the extraction of the Fe-Mn oxyhydroxide fraction. The seawater signal was extracted from the marine sediments applying a sequential leaching technique described in chapter 2. The seawater-derived Fe-Mn oxyhydroxide fractions were extracted using a gentle reducing cocktail (0.05M hydroxylamine hydrochloride – 15 % acetic acid – 0.03M Na-EDTA, buffered to pH 4 with NaOH) for three hours in a shaker at room temperature.

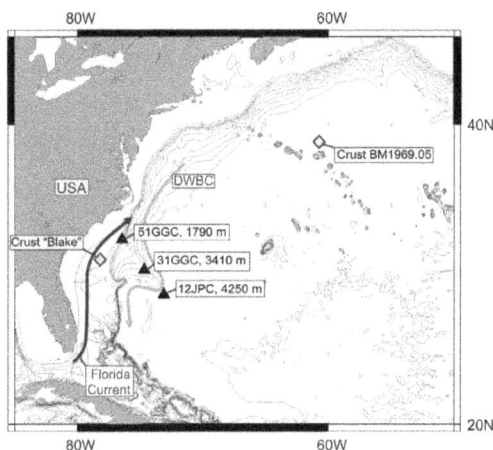

Figure 4.1. Core locations along the Blake Ridge in the western North Atlantic. Trajectories of the schematic flow paths of the Florida current in water depths above 1000 m and the Deep Western Boundary Current (DWBC) transporting North Atlantic Deep Water southward are indicated. Also shown are the sampling locations of ferromanganese crusts BM1969.05 from the New England seamounts (Reynolds et al., 1999; Foster and Vance, 2006), as well as crust "Blake", analysed by Reinolds et al. (1999).

The seawater origin of the extracted Fe-Mn oxyhydroxide fraction was ensured by monitoring Al/Pb elemental ratios and by mass balance calculations (chapter 2). The Fe-Mn oxyhydroxide Pb fraction can be reliably extracted because of its enrichment in the seawater-derived phase compared with the detrital minerals. Authigenic Pb in the sediments along the Blake Ridge accounts for 15-40 % of the entire Pb pool present (*i.e.*, the sum of the Fe-Mn oxyhydroxide and detrital fraction). The

$^{207}Pb/^{206}Pb$ and $^{208}Pb/^{206}Pb$ isotope ratios obtained here for Fe-Mn oxyhydroxide fractions which are supposed to reflect a pure seawater component are in good agreement with published ferromanganese crust data from the western North Atlantic (Reynolds et al., 1999; Foster and Vance, 2006).

Separation and purification of Pb from the Fe-Mn oxyhydroxide matrix followed standard procedures (Lugmair and Galer, 1992). The total procedural Pb blank in the Fe-Mn oxyhydroxide fraction was ~ 1 ng and always below 0.35 % of the total Pb concentration. Measurements of the Pb isotopes were carried out on a Nu Plasma MC-ICPMS at ETH Zürich applying a Tl-doping procedure (Walder and Furuta, 1993; Belshaw et al., 1998) with a Pb/Tl ~ 4 in the measurement solutions. The expected offset relative to TIMS Pb isotope data (cf., Thirlwall, 2002) was accounted for by normalising the respective Pb isotope composition to the triple spike TIMS Pb ratios for NIST SRM981 ($^{206}Pb/^{204}Pb$ = 16.9405, $^{207}Pb/^{204}Pb$ = 15.4963, $^{208}Pb/^{204}Pb$ = 36.7219; (Galer and Abouchami, 1998; Abouchami et al., 1999). Duplicate samples from cores 51GGC and 12JPC were processed through chemistry and measured separately at least six months later. The external reproducibility in the course of this study was ±96 ppm for $^{206}Pb/^{204}Pb$, ±112 ppm for $^{207}Pb/^{204}Pb$, ±142 ppm for $^{208}Pb/^{204}Pb$, ±33 ppm for $^{207}Pb/^{206}Pb$ and ±69 ppm for $^{208}Pb/^{206}Pb$ (2 relative standard deviations). Lead isotope results are displayed in Table 4.1.

Conventional ^{14}C ages published for cores 51GGC (n = 12) and 12JPC (n = 2) (Keigwin, 2004), obtained from planktonic foraminifera, were transformed into calendar years using the marine radiocarbon calibration dataset of Hughen et al. (2004), assuming ΔR = 0. Two ^{14}C ages obtained from planktonic foraminifera in core 12JPC (4250 m) have been published by Keigwin (2004). All radiocarbon and calibrated ages can be found in the Appendix. Besides the two published planktonic radiocarbon ages the radiogenic Pb isotope spike observed in the 51GGC and 12JPC at 11.2 ka BP was used as a third age tie point. Ages between the three tie points have been linearly interpolated. Sedimentation rates between the tie points in core 12JPC were linearly interpolated. For clarity the Pb isotope evolution in core 12JPC is shown plotted both against depth in core in Figure 4.2a, and against the age in Figures 4.4, 4.5 and 4.6. Matching the radiogenic Pb isotope spikes in the two cores is valid for the following reasons: Core 51GGC is chronologically very well constrained (see arrows in Figure 4.3) and both depths were exposed to the same water mass (NADW) during the early Holocene (chapter 3).

Table 4.1. Pb isotope compositions of all analysed Fe-Mn oxyhydroxide fractions

Depth in core (cm)	Calendar Age (ka BP)	$^{206}Pb/^{204}Pb$	$^{207}Pb/^{204}Pb$	$^{208}Pb/^{204}Pb$	$^{207}Pb/^{206}Pb$	$^{208}Pb/^{206}Pb$
		± 0.0016	± 0.0017	± 0.0052	± 0.00003	± 0.00015
Core 51GGC, 1790 m						
40 cm	1.59	19.088	15.690	39.133	0.822	2.050
duplicate		19.087	15.686	39.130	0.822	2.050
60 cm	2.39	19.166	15.695	39.191	0.819	2.045
duplicate		19.134	15.691	39.162	0.820	2.047
160 cm	6.36	19.189	15.697	39.194	0.818	2.043
duplicate		19.167	15.691	39.175	0.819	2.044
220 cm	8.75	19.202	15.696	39.223	0.817	2.043
260 cm	10.34	19.216	15.689	39.289	0.816	2.045
270 cm	10.73	19.219	15.684	39.274	0.816	2.043
290 cm	11.20	19.282	15.692	39.329	0.814	2.040
300 cm	12.06	19.251	15.689	39.283	0.815	2.041
duplicate		19.233	15.686	39.266	0.816	2.042
310 cm	12.89	19.172	15.684	39.192	0.818	2.044
316 cm	13.01	19.144	15.683	39.150	0.819	2.045
duplicate		19.123	15.678	39.131	0.820	2.046
330 cm	13.72	19.123	15.674	39.150	0.820	2.047
duplicate		19.097	15.669	39.106	0.820	2.048
350 cm	15.05	19.101	15.671	39.130	0.820	2.049
duplicate		19.072	15.667	39.076	0.821	2.049
370 cm	17.16	19.093	15.675	39.122	0.821	2.049
duplicate		19.068	15.665	39.074	0.822	2.049
371 cm	17.22	19.069	15.667	39.074	0.822	2.049
duplicate		19.068	15.666	39.074	0.822	2.049
373 cm	17.35	19.061	15.675	39.008	0.822	2.046
378 cm	17.65	19.062	15.674	38.996	0.822	2.046
385 cm	21.36	19.051	15.676	38.965	0.823	2.045
390 cm	22.15	19.040	15.675	38.942	0.823	2.045
duplicate		19.028	15.673	38.930	0.824	2.046
395 cm	22.94	19.019	15.669	39.004	0.824	2.051
400 cm	23.73	18.996	15.671	38.932	0.825	2.050
duplicate		18.980	15.662	38.898	0.825	2.049
405 cm	24.52	18.996	15.669	38.933	0.825	2.049
duplicate		18.993	15.666	38.929	0.825	2.050
412 cm	25.62	19.022	15.669	38.958	0.824	2.048
422 cm	27.20	19.014	15.672	38.996	0.824	2.051

Table 4.1. continued

Depth in core (cm)	Calendar Age (ka BP)	$^{206}Pb/^{204}Pb$	$^{207}Pb/^{204}Pb$	$^{208}Pb/^{204}Pb$	$^{207}Pb/^{206}Pb$	$^{208}Pb/^{206}Pb$
Core 31GGC, 3410 m						
20 cm		19.222	15.686	39.373	0.816	2.048
30 cm		19.276	15.698	39.432	0.814	2.046
72 cm		19.386	15.699	39.558	0.810	2.041
135 cm		19.031	15.668	39.005	0.823	2.050
140 cm		18.997	15.664	38.987	0.825	2.052
150 cm		18.985	15.663	38.973	0.825	2.053
260 cm		18.981	15.667	38.970	0.825	2.053
275 cm		19.034	15.672	39.043	0.823	2.051
330 cm		18.972	15.661	38.934	0.825	2.052
390 cm		18.952	15.658	38.913	0.826	2.053
410 cm		18.946	15.659	38.894	0.827	2.053
420 cm		18.938	15.657	38.882	0.827	2.053
Core 12JCP, 4250 m						
20 cm	4.07	19.210	15.684	39.406	0.816	2.051
duplicate		19.217	15.681	39.402	0.816	2.050
30 cm	6.11	19.226	15.677	39.394	0.815	2.049
50 cm	10.18	19.319	15.680	39.502	0.812	2.045
55 cm	11.20	19.370	15.696	39.594	0.810	2.044
duplicate		19.377	15.693	39.591	0.810	2.043
69 cm	11.91	19.303	15.684	39.467	0.813	2.045
77 cm	12.32	19.141	15.667	39.216	0.818	2.049
85 cm	12.67	19.110	15.670	39.184	0.820	2.050
duplicate		19.112	15.665	39.173	0.820	2.050
102 cm	13.38	19.072	15.658	39.073	0.821	2.049
117 cm	14.00	18.964	15.652	38.966	0.825	2.055
140 cm	14.95	18.982	15.659	38.957	0.825	2.052
156 cm	15.62	18.966	15.657	38.931	0.825	2.053
172 cm	16.28	18.942	15.656	38.921	0.827	2.055
193 cm	17.15	18.956	15.659	38.932	0.826	2.054
203 cm	17.57	18.945	15.647	38.891	0.826	2.053
220 cm	18.27	18.941	15.648	38.879	0.826	2.053
233 cm	18.73	18.933	15.653	38.886	0.827	2.054
duplicate		18.931	15.649	38.874	0.827	2.053
251 cm	19.56	18.941	15.650	38.902	0.826	2.054
263 cm	20.05	18.941	15.652	38.913	0.826	2.054
duplicate		18.944	15.654	38.921	0.826	2.055

4.3   Results

4.3.1   Glacial-interglacial ^{208,207,206}Pb/^{204}Pb isotope evolution

Plots of the seawater ^{206}Pb/^{204}Pb, ^{207}Pb/^{204}Pb and ^{208}Pb/^{204}Pb evolution, as recorded in the deeper cores 12JPC and 31GGC are shown in Figure 4.2 where they are compared with the oxygen isotope stratigraphies of Keigwin (2004). The two cores show similar patterns. Least radiogenic ratios during the LGM (Fig. 4.2, Table 4.1) are followed by fairly constant ^{208,207,206}Pb/^{204}Pb throughout most of the deglaciation (below 117 cm depth in 12JPC; 140 cm in 31GGC). In both cores ^{208,207,206}Pb/^{204}Pb show a pronounced increase after the YD levelling out again to lower ^{208,207,206}Pb/^{204}Pb towards present-day. The termination of the YD cold interval (YD: ~12.9 to 11.5 kyr BP) (Broecker and Denton, 1989; Alley, 2000) occurred immediately before the steep increase towards peak radiogenic seawater Pb isotope compositions in core 12JCP (Fig. 4.2a).

In the deepest core 12JPC, ^{206}Pb/^{204}Pb varies from 18.93 during the LGM to 19.37 after the YD followed by a decline to ratios of 19.21 at present. The lead isotope evolution in core 31GGC shows the same amplitude in ^{206}Pb/^{204}Pb variability, ranging from 18.94 during the LGM to 19.39 after the YD and 19.22 today. Overall, ^{206}Pb/^{204}Pb varies by 2.27 % in 12JPC and 2.32 % in 31GGC between the LGM and the early Holocene. This represents about 55 % of the global ^{206}Pb/^{204}Pb range recorded in surface scrapings of ferromanganese crusts (Frank, 2002) and the maxima in the cores exceed the highest reported ^{206}Pb/^{204}Pb of 19.26 recorded in surface scrapings of ferromanganese crusts from nearby locations (Reynolds et al., 1999).

For ^{207}Pb/^{204}Pb, the range is clearly less wide, ranging from 15.65 to 15.70 in 12JPC and 15.66 to 15.70 in core 31GGC between the LGM and the early Holocene. In contrast to ^{206}Pb/^{204}Pb, a small rise towards more radiogenic ^{207}Pb/^{204}Pb is already found in core 12JPC during the early deglaciation.

The ^{208}Pb/^{204}Pb follow the same trend as ^{206}Pb/^{204}Pb, increasing significantly from LGM ratios of 38.88 (in both cores) to values as high as 39.59 (12JPC) and 39.56 (31GGC) after the YD, with a subsequent decrease to 39.41 (12JPC) and 39.37 (31GGC) in the youngest samples. The largest variation in ^{208}Pb/^{204}Pb is observed in core 12JPC, with an overall range of 1.80 % between the LGM and the early Holocene.

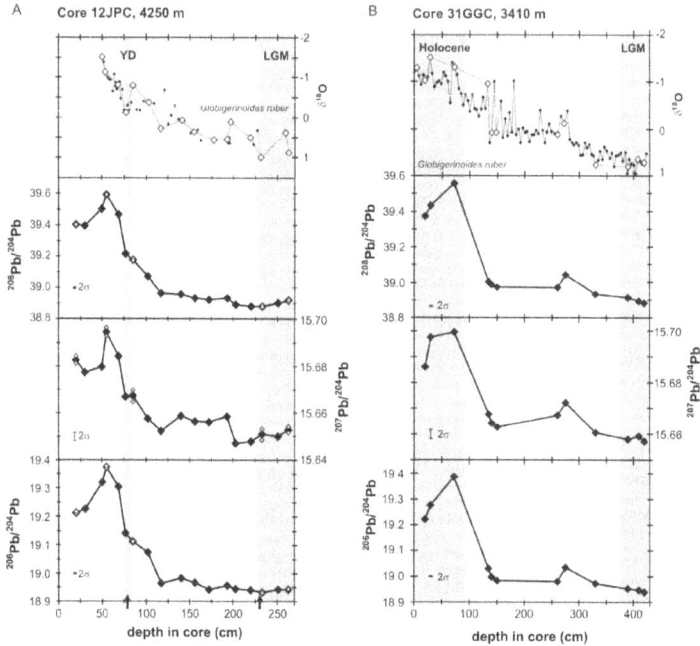

Figure 4.2. Radiogenic Pb isotope results plotted against depth in core for the deep and intermediate cores (a) 12JPC and (b) 31GGC. Oxygen isotope stratigraphy is from Keigwin (2004). Arrows in Figure 4.2a highlight published planktonic radiocarbon ages (Keigwin, 2004). Shaded areas in (a) highlight the Younger Dryas and the LGM. Duplicate samples (white diamonds) analysed for core 12JPC are also shown.

Figure 4.3 illustrates the evolution of $^{208,207,206}Pb/^{204}Pb$ at the location of shallow core 51GGC over the past 27 kyr in the context of the oxygen isotope stratigraphic framework. The overall seawater Pb isotope evolution is similar to the deeper cores, with a considerable shift from low $^{208,207,206}Pb/^{204}Pb$ during the LGM to very radiogenic $^{208,207,206}Pb/^{204}Pb$ after the YD, dropping again within short time to less radiogenic ratios at present-day. The deepest analysed levels in this sediment core predate the LGM (LGM = ~23 to 18 kyr BP) (Yokoyama et al., 2000; Schaefer et al., 2006). Seawater $^{208,207,206}Pb/^{204}Pb$ isotope compositions are lowest at about 24-23 kyr BP. Least radiogenic Pb isotope compositions were recorded before the LGM.

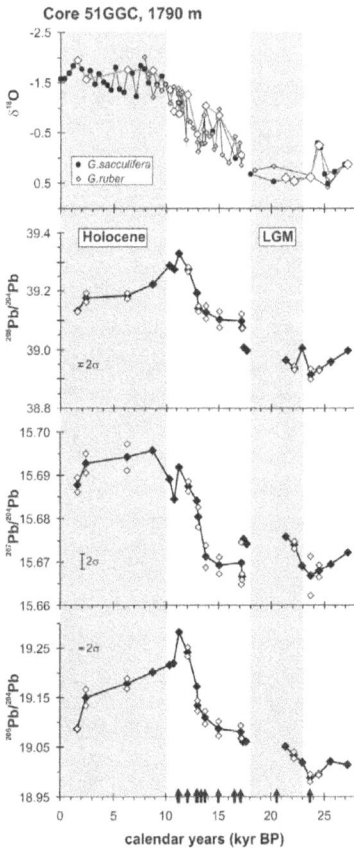

Figure 4.3. Radiogenic Pb isotope results plotted against calendar years for the shallow cores 51GGC. Oxygen isotope stratigraphy is from Keigwin (2004). Arrows along x-axis highlight published planktonic radiocarbon ages (Keigwin, 2004) that have been transformed into calendar years BP (see Appendix). Shaded areas highlight the Holocene and the LGM. Duplicate samples (white diamonds) are also shown.

The lowest $^{206}Pb/^{204}Pb$ of 18.99 is more radiogenic than in the deeper cores (Table 4.1). During the Deglaciation $^{206}Pb/^{204}Pb$ increased slightly but the significant shift towards more radiogenic $^{206}Pb/^{204}Pb$ was not initiated until 13 kyr BP contemporaneous with the onset of the YD. The most radiogenic $^{206}Pb/^{204}Pb$ of 19.28 was recorded at 11.2 ka BP. The radiogenic $^{206}Pb/^{204}Pb$ spike lasted less than at most a few hundred years, and the initially sharply dropping seawater Pb isotope composition continued to decline slowly over the course of the entire Holocene. The total $^{206}Pb/^{204}Pb$ isotope variability is on the order of 1.45 % between the LGM and the early Holocene, which corresponds to about two thirds of the variation seen in the deeper cores.

The trend in $^{207}Pb/^{204}Pb$ isotopes since the last glacial to present-day in core 51GGC differs somewhat from $^{206}Pb/^{204}Pb$, showing a more pronounced minimum before and after the LGM with $^{207}Pb/^{204}Pb$ of 15.67. The LGM itself is not well preserved, however $^{207}Pb/^{204}Pb$ seem to be slightly higher (Figure 4.3). Ratios during the deglaciation remained constant until the beginning of the YD. The following 2 kyr entail the largest shift towards radiogenic $^{207}Pb/^{204}Pb$ of 15.692. In contrast to $^{206}Pb/^{204}Pb$, the most radiogenic $^{207}Pb/^{204}Pb$ of 15.696 is recorded 2.5 kyr later at 8.75 ka BP (Table 4.1). $^{207}Pb/^{204}Pb$ remained constant for the rest of the Holocene apart from the youngest sample. The variability between the LGM and the early Holocene for $^{207}Pb/^{204}Pb$ is 0.14 %, which is only about half the range observed for the glacial-interglacial $^{207}Pb/^{204}Pb$ seawater evolution seen in cores 12JCP and 31GGC.

The $^{208}Pb/^{204}Pb$ seawater curve recorded in 51GGC is roughly similar to the $^{206}Pb/^{204}Pb$ isotope trend. It was, however, more variable during the LGM, and the radiogenic Pb isotope excursion during the early Holocene is less pronounced than for $^{206}Pb/^{204}Pb$. Lowest $^{208}Pb/^{204}Pb$ of 38.91 were recorded at 23.7 ka BP. Throughout the YD, $^{208}Pb/^{204}Pb$ increased until 11.2 ka BP, reaching ratios of 39.33. Ratios then decreased to a modern $^{208}Pb/^{204}Pb$ of 39.13. The overall variability is on the order of 1.07 %, which is slightly more than half the variability seen in the deeper cores.

4.3.2   Combined seawater Pb isotope records

Using the two radiocarbon dates published for core 12JPC, and adjusting the age of the radiogenic $^{206}Pb/^{204}Pb$ spike of core 12JPC to the chronologically well-constrained core 51GGC, it is possible to directly compare the seawater Pb isotope evolution of the deep and shallow western North Atlantic along the Blake Ridge. This has been done in Figure 4.4, where the Pb isotope records of cores 12JPC and 51GGC are shown together. Clearly core 12JPC recorded much larger Pb isotope variations than shallow core 51GGC. Seawater $^{207}Pb/^{206}Pb$ had highest ratios during the LGM ($^{207}Pb/^{206}Pb$ of 0.823 to 0.826), and lowest ratios immediately after the Younger Dryas ($^{207}Pb/^{206}Pb$ of 0.810 to 0.814). Present-day ratios are considerably higher than post-Younger Dryas ratios ($^{207}Pb/^{206}Pb$ of ~0.816 to 0.822). The pattern for $^{208}Pb/^{206}Pb$ resembles that of $^{207}Pb/^{206}Pb$, however the temporal changes are less

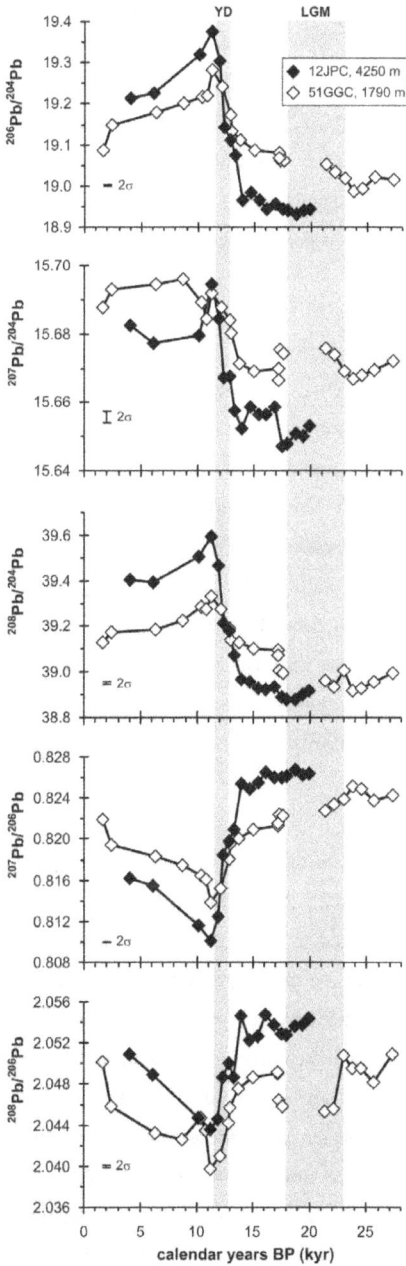

Figure 4.4. Deep and shallow water Pb isotope evolution along the Blake Ridge in the western North Atlantic represented by cores 12JPC and 51GGC. The shaded areas highlight the Younger Dryas (YD) and the LGM.

pronounced than in the ^{207}Pb/^{206}Pb record. Core 51GGC displays a drop at the transition to the LGM at about 23 ka BP, and ^{208}Pb/^{206}Pb during the LGM differs significantly between core 51GGC (^{208}Pb/^{206}Pb ~ 2.046) and 12JPC (^{208}Pb/^{206}Pb ~ 2.054). Ratios converged after the YD (^{208}Pb/^{206}Pb of 2.040 and 2.044 at around 11 ka BP) and subsequently increased towards present-day ratios of ~2.050. The only Pb isotope ratio that does not show a short-term radiogenic spike after the Younger Dryas is ^{207}Pb/^{204}Pb in core 51GGC, which rather reached a plateau after the increase towards the early Holocene.

4.3.3 ^{207}Pb/^{206}Pb and ^{208}Pb/^{206}Pb isotope trends

The small differences observed between the three cores in the glacial-interglacial evolution of the seawater Pb isotope signal are also evident when comparing ^{207}Pb/^{206}Pb and ^{208}Pb/^{206}Pb. In Figure 4.5a, the Fe-Mn oxyhydroxide seawater signals follow roughly linear trends illustrated by the regression lines for the individual cores. The glacial-interglacial ^{207}Pb/^{206}Pb and ^{208}Pb/^{206}Pb of the Fe-Mn oxyhydroxide coatings are plotted together with laser ablation data of Foster and Vance (2006), confirming the good agreement between the two different seawater archives. The glacial-interglacial ^{207}Pb/^{206}Pb and ^{208}Pb/^{206}Pb compositions of deepest core 12JPC matches best those of ferromanganese crust BM1969.05 analysed earlier by Reynolds et al. (1999) and Foster and Vance (2006).

Most prominently, the seawater signal recorded in the deepest core 12JPC displays higher ^{208}Pb/^{206}Pb for a given ^{207}Pb/^{206}Pb compared with the two shallower cores. The spatial and temporal trends can best be illustrated by the time slice reconstructions provided in Figure 4.5b. The regression lines for the three cores indicate convergence for the glacial ^{208}Pb/^{206}Pb and ^{207}Pb/^{206}Pb, and diverge for post-Younger Dryas seawater. Both ^{208}Pb/^{206}Pb and ^{207}Pb/^{206}Pb are highest during the LGM and lowest immediately after the Younger Dryas. Present-day ratios fall between these end-members (see also Table 4.1). Figure 4.5c shows ferromanganese crust data spanning the past 3 Myr (Reynolds et al., 1999), recovered from the Blake Plateau above the Blake Ridge and from the New England Seamounts in the northwest Atlantic (Fig. 4.1). The divergence observed on glacial-interglacial timescales between deep and

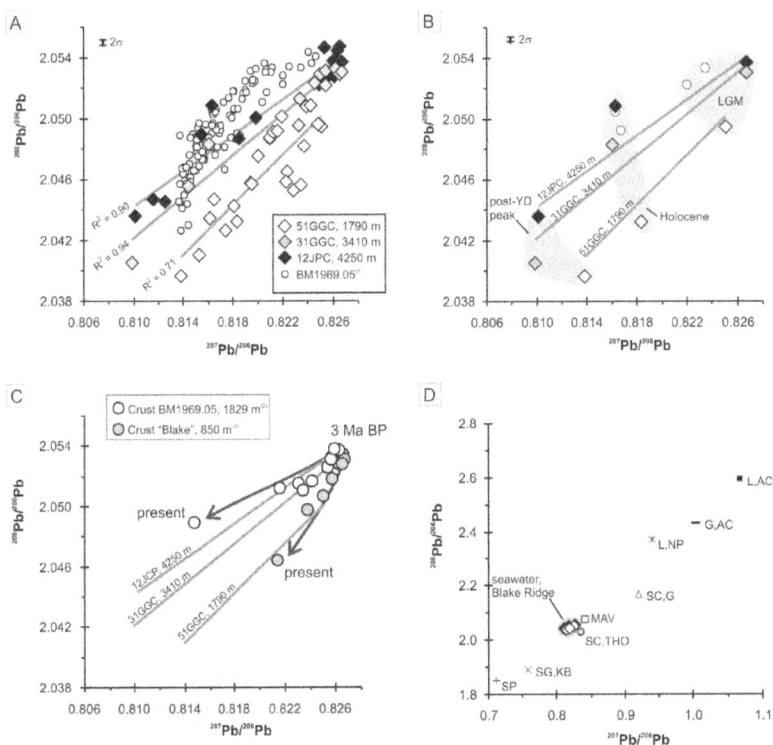

Figure 4.5. **(a)** $^{207}Pb/^{206}Pb$ versus $^{208}Pb/^{206}Pb$ for Fe-Mn oxyhydroxide fractions in the three sediment cores. The Pb isotope signal in the various water depths follows slightly different trends, which is most likely due to Pb contribution from Gulf Stream waters to the Fe-Mn oxyhydroxide fraction in shallow and intermediate cores 51GGC and 31GGC above the Blake Ridge. Linear regression lines and coefficients are displayed for a better distinction of the Pb isotope trends. Also shown in (a) are the laser ablation results of ferromanganese crust BM1969.05 published by Foster and Vance (2006). **(b)** Time slice reconstructions for the LGM, the post-Younger Dryas Pb isotopic excursion and the present situation. Four data points of Foster and Vance (2006) spanning the past 25 kyr are also shown. Symbols are similar to (a). **(c)** Published ferromanganese crust data of Reynolds et al. (1999), using solution chemistry. Note the agreement in Pb isotope trends between crust "Blake", derived from Florida current water depths, and core 51GGC, situated in NADW. **(d)** Possible sources for Pb advected to the Blake Ridge, illustrating the difficulty in depicting a certain source. **SP**: Superior Province, inland Canada, **SG,KG**: Southern Greenland, Ketilidian Belt, **SC,THO**: Southern Canada, Trans-Hudsonian orogen, **MAV**: Mid-Atlantic volcanism, **SC,G**: Southern Canada, Grenvillian belt, **L,NP**: Labrador, Nain Province, **G,AC**: Greenland, Archean craton, **L,AC**: Labrador, early Archean gneisses (all provenance data are compiled in Fagel et al., 2002).

shallow sites along the Blake Ridge (Figs. 4.5a and b) is also recorded on timescales of millions of years (Fig. 4.5c). Highest $^{208}Pb/^{206}Pb$ and $^{207}Pb/^{206}Pb$ in ferromanganese crusts were recorded 3 Myr ago, lowest ratios occurred in surface scrapings, and the

$^{208}Pb/^{206}Pb$ and $^{207}Pb/^{206}Pb$ recorded in Fe-Mn oxyhydroxide coatings along the deeper Blake Ridge (31GGC and 12JPC) during the LGM are identical to ferromanganese crust compositions 3 Myr ago.

Figure 4.5d illustrates the incongruent weathering effect observed for seawater Pb isotopic compositions released in the weathering cycle. The Pb isotope compositions observed on glacial-interglacial timescales could be a mixture of various sources displayed in Figure 4.5d and, based on the $^{207}Pb/^{206}Pb$ and $^{208}Pb/^{206}Pb$ isotope compositions alone, no unequivocal source area can be nominated. Compared with the possible range of source compositions, the Pb isotope variability recorded along the Blake Ridge is comparatively small.

4.3.4   Combined glacial-interglacial Pb and Nd isotope evolution

The seawater Pb isotope compositions recorded in sediment cores along the Blake Ridge in the western North Atlantic reveal a very systematic evolution (Fig. 4.4). Both sediment cores, whether derived from abyssal (4250 m) or from intermediate (1790 m) depth, show the same general trends in the transition from the Last Glacial Maximum (LGM) to the Holocene. The isotopic excursions are most pronounced for $^{206}Pb/^{204}Pb$ and $^{208}Pb/^{204}Pb$, and the amplitude is largest for deep core 12JPC in 4250 m. In this regard it is important to understand the origin of the water masses that ventilated the intermediate and deep Blake Ridge during this time interval. Figure 4.6 illustrates the seawater Pb isotope evolution of deep core 12JPC in context with the Nd isotope composition of the same samples presented in chapter 3. The radiogenic Nd isotope compositions ($\varepsilon_{Nd} \sim -10.5$) in core 12JPC observed until the Younger Dryas reflect the presence of Southern Source Water along the deeper Blake Ridge throughout the deglaciation. The deeper Blake Ridge was only ventilated by Lower NADW ($\varepsilon_{Nd} \sim -13.5$) since the termination of the YD. The $^{207}Pb/^{206}Pb$ isotope record traces the water mass change as suggested by the Nd isotope evolution. Consequently, the LNADW that ventilated the deeper Blake Ridge at 11.2 ka BP supplied the radiogenic Pb pulse towards the western North Atlantic.

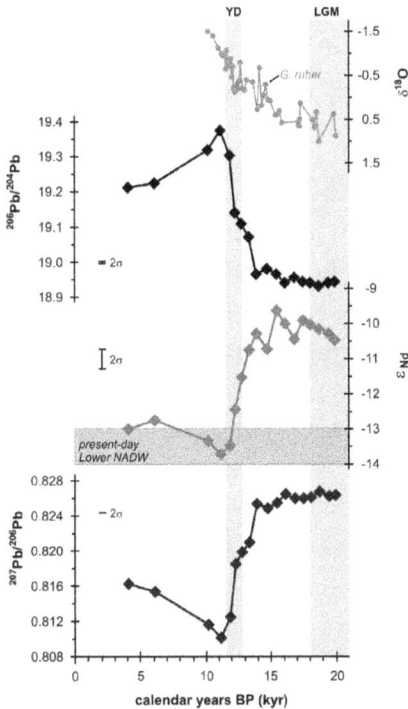

Figure 4.6. Seawater Pb isotope evolution of deep core 12JPC in context with the oxygen isotope stratigraphy (from Keigwin, 2004) and the seawater Nd isotope evolution. The excursion towards the most radiogenic Pb isotope compositions was recorded while LNADW was ventilating the deeper Blake Ridge.

4.4   Discussion

The above observations raise questions regarding the provenance of the seawater Pb preserved in the Fe-Mn oxyhydroxide coatings in Blake Ridge sediments, the glacial-interglacial trends, but most importantly the timing of the radiogenic spike seen in $^{206,207,208}Pb/^{204}Pb$ and $^{207,208}Pb/^{206}Pb$ after the YD in the sediment cores along the Blake Ridge. It has been argued that northwest Atlantic and the Labrador Sea Pb isotopic compositions cannot reflect congruent release from neighbouring continental source rocks else they would be much more depleted in radiogenic lead (see Fagel et al. (2002) for a compilation). Furthermore, the variations observed on glacial-interglacial timescales should be significantly larger due to a wider spread in Pb isotope compositions between the various possible source areas (cf. Fig. 4.5d). Below, isotopic effects introduced from different source provenances will be identified first before the factors are constrained that govern the incongruent release of radiogenic Pb from nearby continental source areas. Finally, the information given by the Pb isotope

103

signal will be used to trace continental runoff patterns from the North American continent in the transition from the LGM to the Holocene.

4.4.1 Florida current contributions to the NADW Pb isotope signal

The linear regression lines included in Figure 4.5a and 4.5b reveal systematic differences in seawater $^{208}Pb/^{206}Pb$ and $^{207}Pb/^{206}Pb$ isotope compositions at different depths along the Blake Ridge. The Pb in deepest core 12JPC has highest $^{208}Pb/^{206}Pb$ for a given $^{207}Pb/^{206}Pb$. Intermediate depth core 31GGC displays slightly lower $^{208}Pb/^{206}Pb$ for a given $^{207}Pb/^{206}Pb$, and core 51GGC exhibits lowest $^{208}Pb/^{206}Pb$ for a given $^{207}Pb/^{206}Pb$. This difference is most pronounced after the YD and today (Fig. 4.5b). This difference can be attributed to contributions from sediments from shallow, near-coastal sites on the upper continental rise of the eastern North America, which were stirred up by the Florida current (Eittreim et al., 1969; Eittreim et al., 1976; McCave, 1986). The effect of surface contributions to the Fe-Mn oxyhydroxide signal was already observed for Nd isotope compositions of the Fe-Mn oxyhydroxide fractions, and was supported by $^{230}Th_{xs}$ results presented from core 51GGC, which suggested high sediment focussing during the Holocene. Thereby oxyhydroxide coatings were redistributed that had acquired their isotopic signature at shallower depths, which have disturbed the original Nd isotope signal of NADW at 1790 m water depth (chapter 3).

Lead export from shallow depths above the Blake Ridge is also evidenced by the Pb isotope evolution trend of ferromanganese crust "Blake" (Fig. 4.5c). Although this crust grew in Gulf Stream waters of the Florida current it closely follows the $^{208}Pb/^{206}Pb$ and $^{207}Pb/^{206}Pb$ trends displayed by shallow core 51GGC, which was deposited in the flow path of upper NADW (Stahr and Sanford, 1999). Therefore, the different $^{208}Pb/^{206}Pb$ and $^{207}Pb/^{206}Pb$ seawater trends displayed by core 51GGC compared with the other cores most likely reflects Pb contributions derived from proximal shallow sites along the south eastern U.S.

4.4.2   A mean ^{207}Pb/^{206}Pb age from Fe-Mn oxyhydroxides coatings

The dramatic increases in ^{208,207,206}Pb/^{204}Pb ratios associated with the onset of the Holocene could be the result of breakdown of accessory minerals and / or release of labile radiogenic Pb released from lattice damaged sites (Hansen and Friderichsen, 1989; Davis and Krogh, 2001; Harlavan and Erel, 2002). If this is the case one can calculate the apparent age of this radiogenic component. A mean "207-206 age" can be calculated by determining a ^{207}Pb*/^{206}Pb* for a zero value of ^{204}Pb/^{206}Pb. In Figure 4.7 seawater ^{204}Pb/^{206}Pb Fe-Mn oxyhydroxide fractions of all three cores define a straight line when plotted against ^{207}Pb/^{206}Pb. The ^{207}Pb/^{206}Pb intercept of 0.0987 for a ^{204}Pb/^{206}Pb of 0 translates into an apparent age of 1.60 Ga. This is an average for the radiogenic component and is similar to that calculated by van der Flierdt et al. (2002) from ferromanganese crusts. This implies derivation of the radiogenic Pb from a mix of the older crustal segments bordering the Labrador Sea and argues against significant delivery of radiogenic Pb from the nearby eastern U.S. represented by the Mid-Paleozoic Appalachians (cf. Gwiazda et al., 1996). Considering the conclusions drawn in section 4.1 it is not considered feasible that all of the lead was only derived from the Labrador Sea.

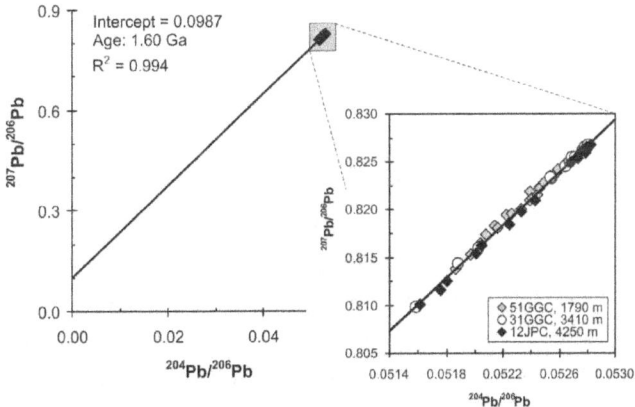

Figure 4.7. Mean ^{207}Pb*/^{206}Pb* age for all Fe-Mn oxyhydroxide fractions calculated analogous to van de Flierdt et al. (2002). The intercept for a ^{204}Pb/^{206}Pb of 0 translates into a mean age of 1.6 Ga.

4.4.3   Reduction of riverine Pb input during LGM?

Another obvious feature in Figures 4.5a and 4.5b is the similarity of $^{208}Pb/^{206}Pb$ and $^{207}Pb/^{206}Pb$ in the deep and shallow cores during the LGM and the subsequent divergence during the transition to the Holocene. Although two water masses of completely different origin prevailed along the Blake Ridge at that time, the seawater $^{208}Pb/^{206}Pb$ and $^{207}Pb/^{206}Pb$ compositions were similar during the LGM. Given the very short residence time of Pb, the question arises from how far away the Pb was supplied. The SSW bathing the lower Blake Ridge during the LGM cannot contain significant amounts of Pb derived from continental source areas in the southern hemisphere due to the short residence time of Pb in seawater. It is unclear at present to what extent SSW was diluted with water of northern origin during advection to the Blake Ridge. Besides North Atlantic waters also Mediterranean Outflow Water may have played a role (e.g., Llave et al., 2006). Clearly the Nd isotope composition of glacial SSW along the Blake Ridge (ε_{Nd} of -10.4) indicates significant mixing with Northern component water considering an ε_{Nd} of original SSW at present-day in the South Atlantic of -6 to -8.5 (Jeandel, 1993; Abouchami et al., 1999; Piotrowski et al., 2004). Hence the similarity in $^{208}Pb/^{206}Pb$ and $^{207}Pb/^{206}Pb$ recorded in the three sediment cores along the Blake Ridge during the LGM may indicate that the SSW Pb isotope signal was dominated by Pb admixed from proximal northern source waters. Generally reduced transfer of continental Pb into the North Atlantic during the LGM, however, could be another explanation.

Klemm et al. (2007) recently argued that the dissolved Pb isotope composition recorded in deep sea hydrogeneous ferromanganese deposits in the major ocean basins is dominated by atmospheric volcanic aerosol Pb thereby forming a background Pb isotope signal to deep waters worldwide. In Figure 4.8, $^{208}Pb/^{206}Pb$ and $^{207}Pb/^{206}Pb$ of the sediment cores along the Blake Ridge are plotted together with the respective Pb isotope compositions of this suggested atmospheric volcanic aerosol end-member (Klemm et al., 2007), represented by Pb isotope compositions of a ferromanganese crust from the western equatorial Pacific (van de Flierdt et al., 2004a). Compared with Holocene seawater Pb isotope compositions along the Blake Ridge, the LGM $^{208}Pb/^{206}Pb$ and $^{207}Pb/^{206}Pb$ in Figure 4.8 appear indeed shifted towards this possible end-member. Should this trend be causally related and not just coincidental, then the riverine Pb input into the North Atlantic must have been significantly reduced, albeit sustained, during the LGM. The reduction of continental Pb input then led to a shift towards the Pb isotope composition of the background atmospheric volcanic aerosols.

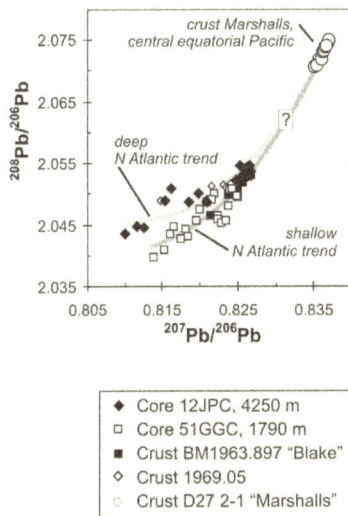

Figure 4.8. Trends in $^{208}Pb/^{206}Pb$ and $^{207}Pb/^{206}Pb$ of cores 12JPC, 51GGC, crust BM1969.05, crust "Blake" (Reynolds et al., 1999) and crust Marshalls from the central equatorial Pacific (van de Flierdt et al., 2004a). The observed convergence of the Pb isotope compositions of Fe-Mn oxyhydroxide coatings could be due to reduced input of continentally-derived Pb into the North Atlantic, leading to a shift toward an atmospheric aerosol input end-member (cf. Klemm et al., 2007).

4.4.4   Agreement between LGM and pre-Pleistocene Pb isotopes?

The North Atlantic ferromanganese crust $^{208}Pb/^{206}Pb$ and $^{207}Pb/^{206}Pb$ compositions predating the onset of major Northern Hemisphere glaciation 2.7 million years ago (Reynolds et al., 1999) have very similar Pb isotope ratios to the LGM data of shallow core 51GGC, and identical ratios to the LGM sections in deeper cores 31GGC and 12JPC. At first sight, these findings appear to be in conflict. One of the major conclusions drawn by Reynolds et al. (1999) was that the rise towards more radiogenic Pb isotope compositions over the past 3 Myr has been caused by an increase in the erosional input into the North Atlantic, caused by the intensification of Northern Hemisphere glaciation after about 2.7 Ma. Bearing this conclusion in mind and considering the suggested decrease in continentally-derived Pb input in the North Atlantic during the LGM (see section 4.4.3) leads to the following implications. The major portion of continental Pb transferred into the western North Atlantic over the past 3 Myr must have been delivered during interglacials and interstadials during or after temporary melting of the Laurentide ice sheet. Retreat of the permafrost allows the climatic northern boundary of chemical weathering to migrate pole-ward. In these areas, vast amounts of fresh rock substrate physically eroded in the preceding glaciation can then be easily chemically weathered. The resulting Pb fluxes from the North American shield to the North Atlantic should be significantly reduced during glacials and were most likely highest during interstadials and incipient interglacials. Therefore the Pb flux from the continents to the North Atlantic should have been by no means at steady state throughout the Pleistocene.

4.4.5   Chemical weathering on the North American continent

In order to understand the glacial-interglacial seawater Pb isotope evolution along the Blake Ridge, the processes controlling the Pb isotope release from fresh rock substrate during incipient and continued chemical weathering need to be considered: (a) The effect of alpha-recoil-damaged mineral lattice sites during incipient chemical weathering of mechanically eroded glacial rock surfaces. (b) The distribution of U, Th and Pb among major and accessory minerals with variable susceptibilities to dissolution during chemical weathering.

Several studies have been published on the preferential release of radiogenic Pb from U-rich minerals such as zircons because Pb loss can affect the reliability of U-Pb geochronology (Hansen and Friderichsen, 1989; Davis and Krogh, 2001; Chen et al., 2002; Romer, 2003). Any U- or Th-rich mineral will suffer increasing degrees of alteration with increasing crystallisation age due to the decay of the respective parent nuclide to the radiogenic Pb daughter. Even zircons, generally considered to be extremely weathering-resistant, will ultimately suffer significant destruction of the crystal lattice in the final stages of metamictization (Murakami et al., 1991). In the course of intense glacial weathering old continental crust is exhumed and crushed. During incipient interglacial conditions the crushed substrate is exposed to chemical weathering, and radiogenic lead loosely bound in damaged lattice sites in minerals can be very efficiently washed out. In Figure 4.4, $^{208}Pb/^{204}Pb$ and $^{206}Pb/^{204}Pb$ display a short-lived extreme radiogenic Pb spike, which is less pronounced for $^{207}Pb/^{204}Pb$. This feature could reflect the above described Pb-loss from alpha-recoil-damaged lattice sites. Lead-207 is the daughter of ^{235}U, which at present-day is close to extinction (0.72 % abundance). Hence the radioactive decay of ^{232}Th to ^{208}Pb, but much more importantly the decay of ^{238}U to ^{206}Pb, will more efficiently destroy crystal lattice sites in continental rocks exposed on the Canadian and Greenland shield because of their much higher abundances and despite their significantly longer half lifes.

The second important parameter governing the release of Pb isotopes during weathering is the susceptibility of the host minerals to weathering. Over the last decade a series of studies were conducted to constrain the release of Pb and its isotopes from granitic soil chronosequences (Erel et al., 1994; Harlavan et al., 1998; Harlavan and Erel, 2002; Erel et al., 2004). These authors found systematic trends during initial and ongoing chemical weathering in these rocks. During early stages of chemical weathering, lower $^{207}Pb/^{206}Pb$ were observed contemporaneously with elevated $^{206}Pb/^{204}Pb$ and/or $^{208}Pb/^{204}Pb$. The offset of the respective Pb isotope ratios from bulk soil compositions decreased with increasing soil age. Additional major and trace element abundances and ratios led these authors to the conclusion that U and Th-rich accessory minerals such as allanite, apatite, sphene, and to a smaller degree biotite and monazite control the Pb isotope signal during the early stages of weathering, and only when these mineral phases are depleted, major mineral constituents such as feldspars dominate the released Pb isotope signal. These authors estimated the time for depletion of the U and Th-rich accessory minerals to be on the order of 10- to 100 kyrs (Erel et al., 1994; Harlavan et al., 1998; Harlavan and Erel, 2002; Erel et al., 2004). It is very feasible that the mechanism of preferential release of radiogenic Pb contributed to the Pb isotope trend observed during the Holocene along the Blake Ridge. The seawater signal along the Blake Ridge displays radiogenic $^{208,207,206}Pb/^{204}Pb$ isotope spikes after the YD, contemporaneous with lowest

^{207}Pb/^{206}Pb, which closely matches the continental weathering trends observed by these authors.

The more efficient release of ^{206}Pb compared with ^{207}Pb was also illustrated by leaching experiments conducted by von Blanckenburg and Nägler (2001). During weak leach experiments using 0.05M HCl on (a) the detrital fraction of ODP core 645 from Baffin Bay, (b) crushed Archean rocks, and (c) Greenland river sand, these authors always extracted a Pb fraction with lower ^{207}Pb/^{206}Pb than the bulk sample ^{207}Pb/^{206}Pb. It remains unclear to date whether the offsets in ^{207}Pb/^{206}Pb between the leachate and the bulk sample were due to the release of loosely bound alpha-recoiled radiogenic Pb or leaching of accessory phases. Overall, the patterns shown in Figure 4.4 suggest a combination of these two processes (i.e., alpha recoil *and* preferential accessory mineral dissolution). However, it remains to be solved which of the two processes in fact controlled the radiogenic Pb isotopic peak that was recorded after the YD.

4.4.6   Persistent continental runoff diversion into the western North Atlantic

Yet unmentioned key players controlling the seawater Pb isotope record along the Blake Ridge are the freshwater pathways supplying continentally-derived Pb to the North Atlantic. The Laurentide ice sheet covered a large portion of North America during the Last Glacial Maximum with a thickness of more than 3 km (Figure 1.4) (Dyke and Prest, 1987; Dyke et al., 2002). It retained a large volume over the entire deglaciation, melting broadly from the southwest towards the northeast (Dyke and Prest, 1987; Licciardi et al., 1998). This melting behaviour over the course of the deglaciation had fundamental consequences on its continental runoff pattern and proglacial lake formation (Marshall and Clarke, 1999). While clear evidence exists for continental runoff directed into the Gulf of Mexico prior to 13,000 years BP (Kennett and Shackleton, 1975; Leventer et al., 1982), no unequivocal evidence was as yet provided for the proposed re-routing of the freshwater drainage from the Laurentide ice sheet in an easterly direction directly into the North Atlantic (Keigwin et al., 1991; deVernal et al., 1996). This lack of evidence is at odds with interpretations about the causes of the inception of the Younger Dryas (YD), which was supposedly initiated by catastrophic meltwater discharges into the North Atlantic (Broecker et al., 1989a;

Teller et al., 2002). Besides the fact that freshwater fluxes during and after the YD were obviously not large enough to significantly affect the surface water oxygen isotope composition in the western North Atlantic, geomorphological information from eastern outlets of proglacial Lake Agassiz also argues against catastrophic meltwater discharges in an easterly direction at the onset of the YD (Broecker, 2006). Acknowledging the information provided by the Gulf of Mexico oxygen isotope data even led to the assumption that meltwater diversion occurred towards the Arctic Ocean (Tarasov and Peltier, 2005; Fisher, 2007). However, geomorphic evidence for freshwater outbursts directly into the North Atlantic has recently been presented (Piper et al., 2007; Rayburn et al., 2007). The question remains whether the scale of such freshwater outbursts was big enough to influence the marine oxygen- or Pb isotope budget.

Figure 4.9 shows the detailed $^{206}Pb/^{204}Pb$ isotope evolution recorded by shallow core 51GGC and deep core 12JPC. The temporal pattern of the evolution of the seawater Pb isotope composition at the Blake Ridge shown in Figure 4.9 contains valuable information about the presence or absence of major meltwater inputs into the North Atlantic. Due to its specific behaviour during incipient chemical weathering, such as the case during the transition from fully glacial to interglacial conditions, large-scale freshwater influx into the North Atlantic should carry a very distinct radiogenic Pb iosotope signal (section 4.4.5). Meltwater pulse 1a (mwp-1a; 14.2 to 13.7 ka BP) (Fairbanks, 1989; Bard et al., 1996) did not leave any detectable changes in the Pb isotopic composition of bottom water at the Blake Ridge (Fig. 4.9). Wherever mwp-1a was centred geographically, a sea level rise on the order of 20 meters within 500 years (Bard et al., 1996) during the deglaciation is such a dramatic event that must have left its imprint on the proximal seawater Pb isotope composition. If mwp-1a had been centered in North America and drainage had been directed into the North Atlantic, this would be reflected in a switch towards more radiogenic compositions. Hence the Pb isotope data confirm earlier suggestions about a non-Atlantic source for the majority of mwp-1a (Clark et al., 1996; Clark et al., 2002). Conversely, the rise towards more radiogenic Pb isotope compositions recorded at the Blake Ridge coincides with the onset of the YD (12.9 ka BP). Keeping in mind the $\delta^{18}O$ minima reported in the Gulf of Mexico prior to 13 ka BP (Leventer et al., 1982; Keigwin et

111

al., 1991) this suggests that indeed freshwater diversions occurred at that time which were linked to a gradual change in drainage pathways from the Gulf of Mexico to the North Atlantic, possibly via the Hudson River and the Gulf of St. Lawrence (Broecker et al., 1989a; Clark et al., 2001; Rayburn et al., 2007). This suggests that the Pb isotopes were a more sensitive proxy for meltwater drainage than oxygen isotopes in the northwest Atlantic because this suggested re-routing of continental runoff did not leave detectable traces in northwest Atlantic oxygen isotope records.

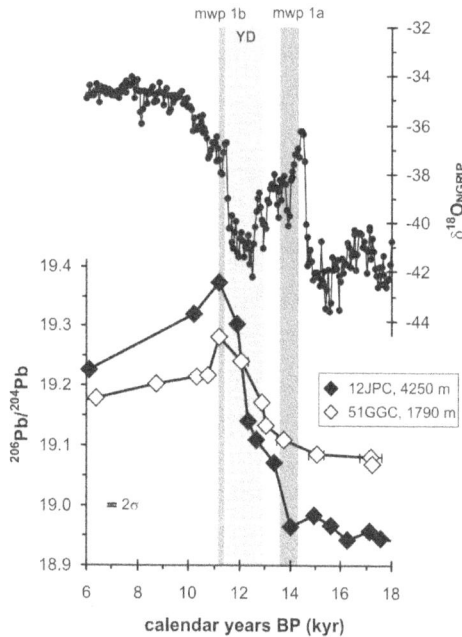

Figure 4.9. Deep and shallow water $^{206}Pb/^{204}Pb$ evolution along the Blake Ridge in the western North Atlantic during the deglaciation and the transition to the Holocene, plotted together with the NGRIP oxygen isotope record of Andersen et al. (2004). The rise towards more radiogenic $^{206}Pb/^{204}Pb$ in intermediate core 51GGC, which was situated in Glacial North Atlantic Intermediate Water during this time interval, initiated at the onset of the Younger Dryas. Note that Pb isotope ratios rise throughout the YD and peak at 11.2 ka BP, contemporaneous with the suggested meltwater pulse 1b (cf. Fairbanks, 1989). The extreme switch in deep core 12JPC also reflects the water mass change from SSW to LNADW (see also section 4.3.4 and Fig. 4.6).

One intriguing aspect about the Pb isotope record in Figure 4.9, however, is the steady rise towards more radiogenic Pb isotope compositions throughout the YD, peaking at

11.2 ka BP, several hundered years after the end of the YD. A change to more radiogenic Pb isotope compositions can only be achieved by increased continental Pb input directly into the western North Atlantic via freshwater. If our Pb isotope records indeed reflect the freshwater fluxes from the North American continent, then freshwater was increasingly diverted into the western North Atlantic in the course of the YD. This observation contrasts with earlier interpretations of short-term re-routing events of continental runoff into the Gulf of Mexico after the YD (e.g. Licciardi et al., 1998; Clark et al., 2001).

The Pb isotope evolution at the Blake Ridge during the deglaciation also challenges Tarasov and Peltier's (2005) glacial system model analysis. In their calculations, the largest total discharge into the Atlantic Ocean (not including the Labrador Sea) during the deglaciation occurred during the mwp-1a interval. In their model, discharge is split between the Hudson River and the St. Lawrence outlet during mwp-1a. The Pb isotope record presented here in contrast suggests no significant direct freshwater influx into the western North Atlantic over this time interval (Fig. 4.9). Conversely, for the time interval that the Pb isotope record presented here suggests the most intense freshwater influx into the North Atlantic (i.e., ~13-11 ka BP), the model results (Tarasov and Peltier, 2005) suggest higher freshwater fluxes into the Arctic Ocean and significantly lower freshwater fluxes into the North Atlantic. Intensified freshwater influx into the Arctic Ocean was, however, not capable of creating the observed patterns seen in Figure 4.9 because the Blake Ridge is located too far away from Arctic input sources to receive a signal in the Pb isotope composition (Frank, 2002).

The most radiogenic Pb isotope compositions shown in Figure 4.9 were recorded in 4250 m water depth and not in core 51GGC in 1790 m water depth. It was argued in section 4.4.1 that contributions from the Florida current influenced the Upper NADW Pb isotope signal (see also Fig. 4.5). Hence, although Florida current contributions offset the Pb isotope compositions of core 51GGC in 1790 m water depth, these contributions were not significant enough to fully obliterate the original Upper NADW radiogenic Pb isotope signal (i.e., the radiogenic spike at 11.2 ka BP is still resolvable in core 51GGC). I suggest that the excursion towards radiogenic Pb isotope

113

compositions in the UNADW during and after the YD were more pronounced than in the deeper NADW.

The Pb isotope data presented here clearly point to a continental source towards the North of the Blake Ridge because the dissolved Pb isotope input was carried within NADW. One major unresolved issue, however, is whether the continental source fed the radiogenic Pb through the Hudson River and the Gulf of St. Lawrence (Piper et al., 2007; Rayburn et al., 2007), or whether the radiogenic Pb was in fact derived from the Labrador Sea. The Nd isotope composition recorded in 4250 m water depth clearly implies a Labrador Sea source for the Nd advected to the Blake Ridge (Fig. 4.6). By inference, one might conclude that the continental Pb must have been derived from the same source. However, because of (a) the incongruent weathering behaviour of Pb during incipient chemical weathering and (b) the much shorter residence time of Pb in seawater we cannot quantify this with the available data.

The seawater Pb isotope results presented here strongly support the notion of gradual and non-catastrophic freshwater drainage diversions from the Gulf of Mexico towards the North Atlantic that persisted throughout the end of the YD. No temporary re-routing of meltwater into the Gulf of Mexico is apparent in this dataset, as opposed to earlier suggestions (Licciardi et al., 1998; Clark et al., 2001). In contrast, the most radiogenic Pb isotope peak at the Blake Ridge is observed at 11.2 ka BP, coinciding - and possibly recording – mwp-1b. Meltwater pulse 1b is the second time interval of excessive freshwater drainage and sea level rise within few hundred years in the course of the deglaciation (Fairbanks, 1989). Its source and origin is yet unconstrained and its existence not entirely proven (Bard et al., 1996) apart from indirect evidence from Atlantic deep-sea coral ventilation ages that indicate an Atlantic origin of mwp-1b (Schroder-Ritzrau et al., 2003).

4.5   Synthesis and Conclusions

The seawater Pb isotope evolution in intermediate and deep waters along the Blake Ridge in the North Atlantic underwent substantial variations since the Last Glacial Maximum. The general pattern seen in all water depths is a shift from unradiogenic Pb isotope compositions to extremely radiogenic compositions during and after the Younger Dryas, followed by a continuous decline towards slightly less radiogenic Pb isotope compositions today. The spatial and temporal Pb isotope trend in the western North Atlantic was governed by both provenance and, more importantly, by climate and the prevailing continental runoff patterns. Our data strongly suggest that fresh rock substrate, physically eroded during the last glacial cycle, incongruently released large quantities of radiogenic Pb during incipient and continued chemical weathering after the retreat of the Laurentide ice sheet and accompanying proglacial lake formation on the North American continent. The very radiogenic Pb isotope compositions recorded along the Blake Ridge after the Younger Dryas reflect most likely the interplay of efficient washout of loosely bound alpha-recoiled radiogenic Pb on the one hand, and the preferential chemical weathering of accessory U- and Th-rich minerals in the latest deglaciation following climate-driven northward migration of the permafrost.

The observed seawater radiogenic Pb isotope excursion recorded in Fe-Mn oxyhydroxide coatings at the Blake Ridge is a direct function of continental runoff reorganisations. Major changes in continental runoff initiated with the onset of the Younger Dryas at around 13 ka BP and culminated at around 11.2 ka BP. Possibly, this marine record provides some of the first direct evidence for enhanced freshwater fluxes into the western North Atlantic during meltwater pulse 1b.

Moreover, the Pb isotope data indicate that the input of continent-derived Pb into the western North Atlantic during the Last Glacial Maximum was significantly reduced. A comparison of the glacial-interglacial seawater Pb isotope trends with ferromanganese crust data from nearby locations spanning the past 3 Myr indicates that any increased input of continent-derived Pb recorded in the western North Atlantic suggested in earlier studies dominantly occurred during interglacials and interstadials, and was probably much lower during glacials.

A radiogenic seawater Pb isotope spike (lasting less than 3 kyr and representing more than half the entire global Pb isotope variability observed from ferromanganese surface scrapings) observed during the early Holocene is a western North Atlantic regional phenomenon. The Pb isotope signal is significantly more radiogenic than any recorded ferromanganese crust Pb isotope composition published to date. This feature illustrates the great potential of past seawater Pb isotope compositions extracted from

Fe-Mn oxyhydroxide fractions in marine sediments to resolve sub-millenial climatic and paleoceanographic excursions, which otherwise cannot be detected.

Acknowledgements

Ben Reynolds, Gavin Foster and Bernard Bourdon are thanked for additional constructive suggestions.

Chapter 5

A first record of glacial-interglacial Hf isotope variations in seawater at sub-millennial resolution

Abstract

The Hf isotope variability of seawater was reconstructed from Blake Ridge sediments for the period from the Last Glacial Maximum (LGM) until present-day on sub-millennial timescales. This study represents the first systematic attempt to extract the Hf signature of seawater from authigenic Fe-Mn oxyhydroxide fractions of marine sediments. Although leaching of detrital minerals apparently obscured the original seawater isotope composition of Hf in some samples, the vast majority of the data seem to reflect a pure Hf isotope signal originating from seawater. Unfortunately no ultimate proof for its seawater-origin can be offered at present due to missing seawater data.

The obtained Hf isotope records show a systematic change from unradiogenic glacial compositions with ε_{Hf} as low as -3.1 in 4250 m water depth and -1.7 in 1790 m water depth to more radiogenic ε_{Hf} today. Today the deep Blake Ridge yields an ε_{Hf} of 1.5, whereas the data from intermediate and shallow depths are in the range of 2.8 ± 0.5. The more radiogenic ε_{Hf} at the shallow sites can probably be attributed to resedimentation of Fe-Mn oxyhydroxide coatings formed in the surface waters upslope analogous to processes identified previously for the Pb and Nd isotope signals from the same Fe-Mn oxyhydroxide fractions.

The seawater ε_{Hf} signal of Southern Source Water (SSW), which prevailed along the deeper Blake Ridge during the LGM and throughout the deglaciation at 4250 m water depth, as inferred from Nd isotopes, was apparently dominated by inputs of northern hemisphere Hf at that time. Alternatively, the SSW that advected to the deeper Blake Ridge during the LGM must have been significantly less radiogenic in ε_{Hf} than suggested by ferromanganese crust data from the Southern Ocean. During the subsequent deglaciation the northern sourced input of continental Hf was apparently less dominant, allowing for the advection of Hf with a more typical SSW isotope composition.

The Hf isotope trends presented here support earlier interpretations of intense mechanical grinding of bulk rock during northern hemisphere glacial weathering. Glacial weathering of continental crust seems indeed to be capable to release enough unradiogenic Hf to shift seawater ε_{Hf} towards significantly less radiogenic compositions. At the same time no clear indication is found that accessory minerals such as apatite, sphene, and allanite have governed the marine Hf isotope signal at any time between the LGM and today. It is suggested that the glacial-interglacial Hf isotope budget in the western North Atlantic is strongly modulated by contributions from (comparatively small) chemical dissolution of zircon, whereas the Pb isotope budget during the deglaciation is rather controlled by other U- and Th-rich accessory mineral phases. Finally, the offset towards more radiogenic ε_{Hf} for a given ε_{Nd} is more likely due to the incongruent release of Hf from continental source rocks than Hf being supplied by hydrothermal activity within the oceans.

5.1 Introduction

The radiogenic isotope composition of Hf in past seawater has attracted a large amount of interest since the advent of multiple-collector inductively coupled plasma mass spectrometry (MC-ICPMS), which facilitated relatively precise measurements (Godfrey et al., 1997; Lee et al., 1999; Piotrowski et al., 2000; David et al., 2001; van de Flierdt et al., 2004b; van de Flierdt et al., 2004c). Hafnium isotope compositions of past seawater possibly yield essential paleoclimatic information especially for the past 1.8 Ma for the reconstruction of ice sheet growth and waning during glacial cycles. Due to its low concentration in seawater, which ranges between 0.1 and 2.4 picomoles per litre, mainly concentration data have so far been published (Godfrey et al., 1996; McKelvey and Orians, 1998; Zimmermann et al., 2004). Almost all of the isotopic information so far has been derived from ferromanganese crusts, which grow very slowly (1-10 mm/Myr) and thus only allow the resolution of variations in the Hf isotope composition of past sweater in the range of 100 kyr to 1 Myr. First direct seawater measurements indicate a strong provinciality of Hf isotopes in the range of ε_{Hf} of -2.9 for Arctic seawater and radiogenic compositions as high as ε_{Hf} of 8.6 in the NW Pacific (Zimmermann et al., 2005).

Based on water column studies (Godfrey et al., 1996) and the geographic variability of Hf isotopes in global deep waters seen between ocean basins based on ferromanganese crust surfaces, the residence time of Hf in seawater has been estimated to be on the order of 1,500 to 2,000 years, which is slightly higher than that of Nd (Godfrey et al., 1996; Lee et al., 1999; David et al., 2001; Frank, 2002). The provinciality in $^{176}Hf/^{177}Hf$ compositions between ocean basins is smaller than that of $^{143}Nd/^{144}Nd$ compositions (Godfrey et al., 1997; David et al., 2001). While a generally coupled behaviour between Hf and Nd isotopes has been observed (Albarède et al., 1998), exceptions have been reported in cases where the two isotope systems have obviously been controlled by different processes (Piotrowski et al., 2000; van de Flierdt et al., 2004b).

The very first Hf isotopic studies of ferromanganese nodules provided evidence that $^{176}Hf/^{177}Hf$ is offset to more radiogenic values for a given $^{143}Nd/^{144}Nd$ in seawater compared with bulk crustal or mantle rocks (White et al., 1986; Albarède et al., 1998; Vervoort et al., 1999). Two main hypotheses for this behaviour have been proposed: One possibility is that hydrothermal Hf is admixed into seawater (White et al., 1986), in contrast to Nd, which is efficiently scavenged around hydrothermal vents (Halliday et al., 1992). Very little is known about the behaviour of Hf in hydrothermal vents systems. However, Bau (1996) argued that the scavenging chemistry was completely different from the REEs. The second possibility is that Hf is released incongruently during breakdown of the continental crust by continental weathering. More than half the continental budget of Hf is stored in the mineral zircon. In contrast, zircons contribute only a small proportion of the REE budget of the crust. Zircon has extremely low Lu/Hf, leading to a major fraction of the continents' unradiogenic Hf being locked up in a relatively indestructible phase. Weathering of the continents produces runoff that is offset to radiogenic Hf for a given Nd isotopic composition, because it is zircon-free (Piotrowski et al., 2000; van de Flierdt et al., 2002). These two models have fundamentally different implications for the Hf budget in seawater. The first implies a hydrothermal dominance whereas the second implies a weathering signal.

The importance of hydrothermal Hf inputs into the oceans is, however, questionable because unambiguous hydrothermal Hf isotope compositions have so far only be measured in ferromanganese crusts immediately at hydrothermal vent sites (Frank et al., 2006). Conversely, ferromanganese crust data have been presented from relatively proximal positions relative to hydrothermal vent sites containing hydrothermal Pb but no hydrothermal Hf (van de Flierdt et al., 2004c). Besides, it is not clear at present if Hf is released to seawater at all due to the lack of data on hydrothermal vent fluids and hydrothermal plumes.

Recently, a first study on the Hf isotope composition in river waters provided the first direct evidence that dissolved ^{176}Hf/^{177}Hf in rivers can be significantly more radiogenic than the corresponding suspended load (Bayon et al., 2006). These authors attributed the observed offset to the preferential release of radiogenic Hf from Lu-rich phases such as apatite, sphene and allanite during chemical weathering, whereas minerals hosting unradiogenic Hf such as zircons are much more resistant against chemical dissolution during weathering.

Bau and Koschinsky (2006) posed the question whether any continental Hf can be transferred from the continents to the ocean via rivers at all in view of its chemical behaviour and speciation. According to these authors Hf is only associated with Fe oxides on which it forms surface precipitates that do not exchange with seawater. Accordingly, riverine terrigenous Hf should not be supplied to the oceans in solution but would be quantitatively removed when Hf-carrying colloids aggregate in the estuaries and coastal waters, where they are subsequently deposited in near-coastal sediments (Bau and Koschinsky, 2006). This process would be capable of producing very radiogenic Hf isotope compositions relative to a given ^{143}Nd/^{144}Nd in seawater, reflecting a dominating hydrothermal signal in seawater due to the lack of continental input. It remains to be demonstrated, however, if riverine Hf is really quantitatively removed in the estuaries and if there is any contribution from hydrothermal solutions at all.

This study seeks to shed new light on the issue of the sources of Hf in seawater by using a so far unused paleoceanographic archive for the reconstruction of the Hf isotope composition of seawater on much shorter timescales (sub-millennial as opposed to several 100 kyrs). Seawater-derived Hf, whether colloidal or truly dissolved, is also incorporated into Fe-Mn oxyhydroxide fractions in marine pelagic sediments. In chapter 2 it was demonstrated that seawater-derived Nd, Pb and Th isotope compositions can be reliably extracted from these Fe-Mn oxyhydroxide coatings. The extraction of a pure seawater-derived Hf fraction from sediments is less straightforward. The results shown below will demonstrate both the potential but also limitations associated with the extraction of a pure seawater Hf signal from marine drift sediments.

5.2   Material and Methods

Three sediment cores spanning the interval from the LGM to present-day and recovered during KNR140 cruise on a transect along the Blake Ridge in the western North Atlantic were selected for Hf isotope analyses. Cores 51GGC from 1790 m water depth, 31GGC from 3410 m depth and 12JPC from 4250 m depth were sampled to reconstruct the seawater Hf isotope evolution in the shallow, intermediate and deep western North Atlantic over the past 30 kyr. In addition, the LGM sections of KNR140 cores 50GGC (1903 m depth), 01JPC (2243 m depth), 02JPC (2394 m depth), 43GGC (2590 m depth) and the LGM section of site 1054A (1300 m depth) drilled during ODP cruise 172 were analysed. The method to extract the seawater-derived Hf isotope signal through sequential reductive leaching followed that described in chapter 2. Because Hf is a highly particle-reactive element, 0.03M Na-EDTA was admixed to the reductive leach solution as a complexing reagent to avoid re-adsorption of Hf leached during Fe-Mn oxyhydroxide dissolution. For a set of eight samples from cores 51GGC and 12JPC the corresponding detrital fraction was completely dissolved and analysed as well (see chapter 2 for analytical procedures).

Separation and purification of the Hf fraction from the Fe-Mn oxyhydroxide and the detrital fractions followed the method of Münker et al. (2001). The total procedural Hf blank of the Fe-Mn oxyhydroxide analyses was below 30 pg (below 0.3 % of the total Hf concentration) and below 20 pg for the detrital fraction in the sediment, hence negligible. Hafnium isotope analyses were conducted on a Nu Plasma MC-ICPMS at ETH Zürich. Measured Hf isotope compositions were normalised to a ^{179}Hf/^{177}Hf of 0.7325 to correct for instrumental mass bias. Mass interferences for various Hf isotopes were monitored by measuring the intensities of ^{172}Yb, ^{175}Lu and ^{182}W during every Hf isotope measurement. Hafnium isotope compositions were normalised to a ^{176}Hf/^{177}Hf of 0.282160 for the Hf standard JMC475 (Nowell et al., 1998). The external reproducibility in the course of this study was ±0.49 ε_{Hf} for ^{176}Hf/^{177}Hf (2σ, n=61). Detailed results of the Hf isotope analyses can be found in Table 5.1 and 5.3.

The timescale used for cores 12JPC and 51GGC is identical to that described in chapter 4. For clarity the ^{176}Hf/^{177}Hf is converted into epsilon notation relative to a chondritic uniform reservoir (CHUR):

$$\varepsilon_{Hf} = \left[\frac{^{176}Hf/^{177}Hf_{sample}}{^{176}Hf/^{177}Hf_{CHUR}} - 1 \right] \times 10^4$$

^{176}Hf/^{177}Hf $_{CHUR}$ = 0.282772 (Blichert-Toft and Albarède, 1997)

Table 5.1. Hf and Sr isotope compositions of all analysed Fe-Mn oxyhydroxide fractions

Depth in core (cm)	Calendar Age (ka BP)	^{176}Hf/^{177}Hf	ε_{Hf}	^{87}Sr/^{86}Sr
		± internal error (2σ)	± applicable error (2σ)	± internal error (2σ)
KNR140, Core 51GGC, 1790 m				
40 cm	1.59	0.282860 ± 10	3.1 ± 0.50	0.709614 ± 14
duplicate		0.282815 ± 29	1.5 ± 1.03	0.711523 ± 12
60 cm	2.39	0.282872 ± 14	3.5 ± 0.50	0.711042 ± 19
duplicate		0.282834 ± 30	2.2 ± 1.07	0.709181 ± 12
160 cm	6.36	0.282835 ± 11	2.2 ± 0.50	0.710655 ± 15
duplicate		0.282863 ± 27	3.2 ± 0.95	0.710463 ± 15
220 cm	8.75	0.282829 ± 10	2.0 ± 0.50	0.709958 ± 10
260 cm	10.34	0.282799 ± 9	0.9 ± 0.50	0.710360 ± 18
270 cm	10.73	0.282868 ± 14	3.4 ± 0.50	0.709576 ± 12
290 cm	11.20	0.282876 ± 9	3.7 ± 0.50	0.711359 ± 15
300 cm	12.06	0.282870 ± 9	3.5 ± 0.50	0.712030 ± 14
duplicate		0.282825 ± 18	1.9 ± 0.64	0.712866 ± 12
310 cm	12.89	0.282930 ± 13	5.6 ± 0.50	0.712600 ± 20
330 cm	13.01	0.282960 ± 15	6.6 ± 0.51	0.714043 ± 15
duplicate		0.282888 ± 11	4.1 ± 0.50	0.711554 ± 11
350 cm	15.05	0.282939 ± 10	5.9 ± 0.50	0.713008 ± 15
duplicate		0.282870 ± 9	3.5 ± 0.50	0.709773 ± 13
370 cm	17.16	0.282948 ± 10	6.2 ± 0.50	0.713194 ± 13
duplicate		0.282850 ± 11	2.7 ± 0.50	0.714968 ± 12
371 cm	17.22	0.282930 ± 6	5.6 ± 0.50	0.711221 ± 11
duplicate		0.282835 ± 11	2.2 ± 0.50	0.713560 ± 17
373 cm	17.35	0.282840 ± 7	2.4 ± 0.50	0.709537 ± 12
378 cm	17.65	0.282867 ± 10	3.4 ± 0.50	0.709491 ± 11
385 cm	21.36	0.282805 ± 10	1.2 ± 0.50	0.709325 ± 12
390 cm	22.15	0.282797 ± 11	0.9 ± 0.50	0.709104 ± 9
duplicate		0.282763 ± 13	-0.3 ± 0.50	0.708940 ± 14
395 cm	22.94	0.282774 ± 15	0.1 ± 0.54	0.713242 ± 19
400 cm	23.73	0.282734 ± 8	-1.4 ± 0.50	0.710821 ± 15
duplicate		0.282741 ± 15	-1.1 ± 0.54	0.709835 ± 10
405 cm	24.52	0.282735 ± 7	-1.3 ± 0.50	0.713064 ± 13
duplicate		0.282713 ± 16	-2.1 ± 0.56	0.716739 ± 14
412 cm	25.62	0.282765 ± 9	-0.3 ± 0.50	0.709755 ± 12
422 cm	27.20	0.282911 ± 9	4.9 ± 0.50	0.710332 ± 14

Table 5.1. continued

Depth in core (cm)	Calendar Age (ka BP)	^{176}Hf/^{177}Hf	ϵ_{Hf}	^{87}Sr/^{86}Sr
		± internal error (2σ)	± applicable error (2σ)	± internal error (2σ)
KNR140, Core 31GGC, 3410 m				
20 cm		0.282850 ± 13	2.8 ± 0.50	0.712302 ± 12
30 cm		0.282849 ± 11	2.7 ± 0.50	0.715800 ± 23
72 cm		0.282797 ± 10	0.9 ± 0.50	0.714190 ± 26
135 cm		0.282862 ± 8	3.2 ± 0.50	0.713987 ± 20
140 cm		0.282886 ± 8	4.0 ± 0.50	0.714633 ± 17
150 cm		0.282893 ± 9	4.3 ± 0.50	0.714899 ± 26
260 cm		0.282940 ± 9	5.9 ± 0.50	0.714134 ± 23
275 cm		0.282921 ± 5	5.3 ± 0.50	0.714300 ± 21
330 cm		0.282963 ± 7	6.8 ± 0.50	0.714758 ± 28
390 cm		0.282906 ± 9	4.8 ± 0.50	
410 cm		0.282891 ± 18	4.2 ± 0.64	0.713070 ± 13
420 cm		0.282855 ± 6	3.0 ± 0.50	0.714393 ± 35
KNR140, Core 12JCP, 4250 m				
20 cm	4.07	0.282815 ± 6	1.5 ± 0.50	0.716419 ± 16
duplicate		0.282814 ± 7	1.5 ± 0.50	0.721992 ± 12
30 cm	6.11	0.282829 ± 6	2.0 ± 0.50	0.715410 ± 17
50 cm	10.18	0.282777 ± 7	0.2 ± 0.50	0.716365 ± 20
55 cm	11.20	0.282771 ± 7	0.0 ± 0.50	0.715589 ± 16
duplicate		0.282780 ± 11	0.3 ± 0.50	0.715004 ± 11
69 cm	11.91	0.282766 ± 5	-0.2 ± 0.50	0.715460 ± 13
77 cm	12.32	0.282775 ± 6	0.1 ± 0.50	0.715112 ± 18
85 cm	12.67	0.282788 ± 8	0.6 ± 0.50	0.715587 ± 17
duplicate		0.282790 ± 9	0.6 ± 0.50	0.713909 ± 11
102 cm	13.38	0.282840 ± 6	2.4 ± 0.50	0.713444 ± 15
117 cm	14.00	0.282752 ± 7	-0.7 ± 0.50	0.713071 ± 16
140 cm	14.95	0.282786 ± 5	0.5 ± 0.50	0.713348 ± 20
156 cm	15.62	0.282822 ± 9	1.8 ± 0.50	0.711928 ± 15
172 cm	16.28	0.282781 ± 6	0.3 ± 0.50	0.712103 ± 19
193 cm	17.15	0.282752 ± 7	-0.7 ± 0.50	0.715013 ± 14
203 cm	17.57	0.282764 ± 7	-0.3 ± 0.50	0.712250 ± 16
220 cm	18.27	0.282727 ± 9	-1.6 ± 0.50	0.713014 ± 14
233 cm	18.73	0.282697 ± 6	-2.6 ± 0.50	0.715062 ± 20
duplicate		0.282671 ± 8	-3.6 ± 0.50	0.721761 ± 16
251 cm	19.56	0.282720 ± 6	-1.8 ± 0.50	0.712150 ± 20
263 cm	20.05	0.282682 ± 15	-3.2 ± 0.54	0.714950 ± 16
duplicate		0.282702 ± 9	-2.5 ± 0.50	0.712391 ± 13

Table 5.1. continued

Depth in core (cm)	$^{176}Hf/^{177}Hf$	ε_{Hf}	$^{87}Sr/^{86}Sr$
	± internal error (2σ)	± applicable error (2σ)	± internal error (2σ)

Additional Last Glacial Maximum sections

ODP 172 1054A, 1300 m			
492 cm	0.282831 ± 38	2.1 ± 1.36	0.709394 ± 13
502 cm	0.282836 ± 11	2.3 ± 0.50	0.709330 ± 11
512 cm	0.282879 ± 16	3.8 ± 0.57	0.709307 ± 13
KNR140, 52GGC, 1710 m			
395 cm	0.282838 ± 14	2.3 ± 0.50	0.709961 ± 13
405 cm	0.283059 ± 11	10.1 ± 0.50	0.710818 ± 13
385 cm	0.282811 ± 11	1.4 ± 0.50	0.710019 ± 12
KNR140, 50GGC, 1903 m			
260 cm	0.282831 ± 12	2.1 ± 0.50	0.710181 ± 13
270 cm	0.282756 ± 12	-0.6 ± 0.50	0.710787 ± 14
KNR140, 01JPC, 2243 m			
167 cm	0.282740 ± 11	-1.1 ± 0.50	0.710255 ± 15
183 cm	0.282739 ± 28	-1.2 ± 1.00	0.710726 ± 16
KNR140, 02JPC, 2394 m			
21 cm	0.282755 ± 15	-0.6 ± 0.50	0.711316 ± 19
28 cm	0.282834 ± 25	2.2 ± 0.89	0.711173 ± 21
KNR140, 43GGC, 2590 m			
182 cm	0.282700 ± 13	-2.6 ± 0.50	0.711506 ± 14
187 cm	0.282704 ± 7	-2.4 ± 0.50	0.711412 ± 16

A series of duplicate measurements were carried out to evaluate the reproducibility of the Hf isotope extraction during reductive leaching. In Table 5.1 and Figure 5.4 the individual results are displayed. For the reconstruction of the Hf isotope evolution of cores 51GGC and 12JPC the mean Hf isotope composition for every sampled depth was used.

5.3 Results

The results section is subdivided into two subsections. First, results are presented to characterise the leached Hf in the Fe-Mn oxyhydroxide fraction. Second, the Hf isotope results extracted through reductive leaching will be presented and discussed together with the Pb and Nd isotope results presented in chapters 3 and 4.

5.3.1 Grain-size effects

In order to constrain the origin of the Hf separated from the oxyhydroxide fraction of the sediments, different grain-size fractions were separated from the bulk sediment, leached, and aliquoted for Hf and Sr analyses from three samples of core 51GGC (1790 m water depth). Figure 5.1 shows the ε_{Hf} data and the corresponding $^{87}Sr/^{86}Sr$ for the Fe-Mn oxyhydroxide leaches from the bulk, fine (< 63 µm) and coarse (> 63 µm) fractions from one Holocene and two LGM sections of core 51GGC. While the Hf isotope compositions extracted from the bulk and the fine fractions are identical within error for all three sampled depths, the Hf isotope compositions of the coarse fractions are significantly different (Fig. 5.1). Furthermore the ε_{Hf} values of bulk samples 51GGC-395 cm and 51GGC-412 cm representing the LGM, are essentially identical whereas the ε_{Hf} values extracted from the coarse fractions differ by 5.3 ε_{Hf} (Fig. 5.1). The scatter observed for the coarse grain size separates does not correlate with Sr isotope compositions. Although every $^{87}Sr/^{86}Sr$ presented in Figure 5.1 is offset from seawater compositions (i.e., $^{87}Sr/^{86}Sr$ = 0.70918) to varying degrees, the trends observed between ε_{Hf} and $^{87}Sr/^{86}Sr$ are not coupled. This indicates that $^{87}Sr/^{86}Sr$ is also not a reliable indicator for the detection of detrital contributions to the Hf isotope signal.

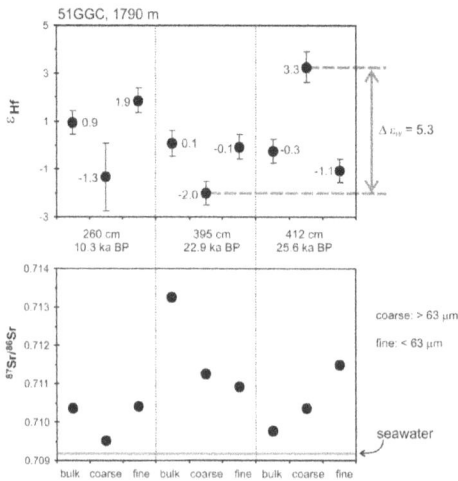

Figure 5.1. Hafnium and Sr isotope results for grain size separates. The Fe-Mn oxyhydroxide fraction was extracted and measured for the bulk, fine (< 63 µm) and coarse (> 63 µm) fraction of three sediment samples in core 51GGC in 1790 m water depth. Numbers in Hf isotope plot refer to ε_{Hf} measured for the respective Fe-Mn oxyhydroxide fraction.

The analyses of the Fe-Mn oxyhydroxide fractions extracted from the coarse sediment fractions resulted in highly variable Hf isotope compositions accompanied by large analytical errors due to low Hf concentrations, which is the reason why only bulk sediment leaches were used for the analyses of this study. Whether the Hf extracted from the Fe-Mn oxyhydroxide fraction of the bulk sediment is purely seawater-derived or partially released during leaching of detrital clays will be discussed in section 5.4.1.

5.3.2   Elemental ratios

In chapter 2, Al/Nd, Al/Pb and Al/Th were presented for two successive Fe-Mn oxyhydroxide leaches together with the respective elemental ratios for the detrital phase of the same sediment samples. Generally, Al/(Nd, Pb, Th) were lowest for the first Fe-Mn oxyhydroxide coating fractions after three hours of leaching, slightly elevated for a second aliquot extracted after another 24 hours of reductive leaching, and orders of magnitude higher for the detrital fraction. The compositional differences observed during these experiments reflect the enrichment of trace metals Nd, Pb and Th in the Fe-Mn oxyhydroxide fraction relative to Al. This is the result of preferential incorporation (sorption and co-precipitation) of seawater-derived particle reactive trace metals into the authigenic Fe-Mn oxyhydroxides. The trace metal enrichment served as evidence in chapter 2 to support the seawater origin of the extracted phase.

Pursuing the same approach for Al/Hf reveals a similar though less pronounced behaviour of Hf. As with Al/(Nd, Pb, Th), the first Fe-Mn oxyhydroxide leach displays lower Al/Hf than the respective detrital fraction (Fig. 5.2, Table 5.2), whereas the second Fe-Mn oxyhydroxide fraction is in the same range of Al/Hf obtained for the detrital fraction. The Al/Hf of the first leaches are in agreement with average Pacific ferromanganese crust ratios with a mean Al/Hf of 1445 (dashed horizontal line in Fig. 5.2; Table 5.2) (Hein et al., 1999).

127

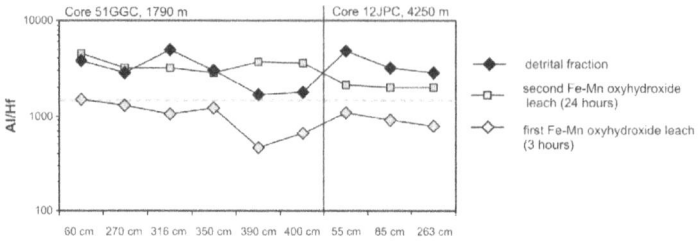

Figure 5.2. Aluminium/hafnium elemental ratios for three different chemical phases of samples presented in Table 5.2. Aluminium/hafnium for the leached coatings that were used for the isotope analyses (light grey diamonds) average to 989 and are lower than average hydrogenetic ferromanganese crusts in the equatorial Central and South Pacific yielding an mean Al/Hf of 1445 (gray horizontal dashed lines; see also Table 5.2). Elemental ratios of a subsequent leach however fall in the range of Al/Hf of the detrital fractions. Note the logarithmic y-axis.

The less pronounced differences in Al/Hf between the different leaches and the detrital fraction can be attributed to the very low abundances of Hf in the first and second Fe-Mn oxyhydroxide coating fractions (Table 5.2). Hafnium concentrations in the first Fe-Mn oxyhydroxide leach fraction range from 24 to 100 ng per gram of leached sediment, orders of magnitude lower than Nd and Pb in the same fractions (Table 5.2). This is also reflected in differences in Nd/Hf and Pb/Hf between the three different phases presented in Table 5.2. Hafnium is clearly depleted in the Fe-Mn oxyhydroxide fraction relative to Nd and Pb if compared with the respective Nd/Hf and Pb/Hf in the detrital fraction. This is consistent with the Hf/Nd and Hf/Pb ratios of ferromanganese crusts (Hein et al., 1999).

128

Table 5.2. Al, Nd, Hf and Pb concentrations of ferromanganese crusts and sediment fractions

	Al(mg/g)[1,2]	Hf (ppm)	Nd (ppm)	Pb (ppm)	Al/Hf	Nd/Hf	Pb/Hf
Central Equatorial Pacific [1]							
Marshall Is.	11.9	8.4	170	1799	1417	20	214
Johnston I.	14.2	10.5	210	1871	1352	20	178
South Pacific [1]	15.8	10.1	226	741	1567	22	73
				Average:	1445	21	155

THIS STUDY	Al (μg/g)[3]	Hf (μg/g)[3]	Nd (μg/g)[3]	Pb (μg/g)[3]	Al/Hf	Nd/Hf	Pb/Hf
First Fe-Mn oxyhydroxide leach (3 hours)							
51GGC - 60 cm	36.1	0.024	3.52	1.37	1501	146	57
51GGC - 270 cm	49.0	0.038	4.95	2.18	1292	130	58
51GGC - 316 cm	24.7	0.024	2.50	1.14	1034	104	48
51GGC - 350 cm	50.4	0.041	3.89	2.63	1228	95	64
51GGC - 390 cm	11.9	0.026	2.31	1.10	459	89	42
51GGC - 400 cm	31.0	0.048	2.64	1.79	649	55	37
12JPC - 55 cm	97.7	0.091	4.19	6.35	1069	46	69
12JPC - 85 cm	82.2	0.091	4.68	6.12	899	51	67
12JPC - 263 cm	77.0	0.100	4.25	4.32	772	43	43
				Average:	989	84	54
Second Fe-Mn oxyhydroxide leach (24 hours)							
51GGC - 60 cm	85	0.019	1.56	0.74	4479	82	39
51GGC - 270 cm	103	0.032	2.90	1.53	3210	90	48
51GGC - 316 cm	48	0.015	0.81	0.97	3211	54	66
51GGC - 350 cm	87	0.031	2.12	2.16	2800	68	69
51GGC - 390 cm	53	0.015	0.61	0.39	3639	42	27
51GGC - 400 cm	80	0.023	0.75	0.43	3556	33	19
12JPC - 55 cm	116	0.055	1.83	2.09	2102	33	38
12JPC - 85 cm	122	0.061	2.41	2.78	1988	39	45
12JPC - 263 cm	146	0.073	2.70	2.21	1997	37	30
				Average:	2998	53	42
Detrital fraction							
51GGC - 60 cm	7640	2.00	17.30	6.41	3818	8.6	3.2
51GGC - 270 cm	11054	3.87	26.63	11.89	2853	6.9	3.1
51GGC - 316 cm	6408	1.31	5.97	2.70	4896	4.6	2.1
51GGC - 350 cm	11375	3.75	27.35	9.40	3036	7.3	2.5
51GGC - 390 cm	6826	4.09	15.34	6.05	1668	3.7	1.5
51GGC - 400 cm	10190	5.72	24.10	8.04	1782	4.2	1.4
12JPC - 55 cm	14637	3.07	24.83	10.00	4770	8.1	3.3
12JPC - 85 cm	12956	4.07	40.82	12.79	3182	10.0	3.1
12JPC - 263 cm	13369	4.78	39.47	15.41	2797	8.3	3.2
				Average:	3200	6.9	2.6

[1] Literature data from Hein et al. (1999)

[2] Where possible Al concentrations were calculated on a loss-on-ignition free base.

[3] Concentration data given for sediments analysed in this study are normalised to μg per gram of raw sediment weighed in.

5.3.3　Mass balance calculations

An additional tool for the determination of the potential corruption of a seawater composition is mass balance. Based on the $^{87}Sr/^{86}Sr$ ratios of the detrital fraction, present-day seawater ($^{87}Sr/^{86}Sr = 0.70918$) and the Fe-Mn oxyhydroxide fractions, the potential contributions of the detrital phase to the Nd and Pb isotopic compositions of the extracted Fe-Mn oxyhydroxide coating can be determined if one knows the Sr, Nd and Pb concentrations in the two phases (Chapter 2). The same calculations can be carried out for the $^{176}Hf/^{177}Hf$ in the coatings, and a theoretical true seawater Hf isotope composition determined. The individual calculations are shown in chapter 2 (Equations 1 to 8) and only the results of the calculations applied to ε_{Hf} are shown in Table 5.3. The measured and calculated ε_{Hf} for the Fe-Mn oxyhydroxide coatings and the calculated seawater ε_{Hf} are also shown in Figure 5.3, in addition to the Hf isotope composition of the respective detrital fraction.

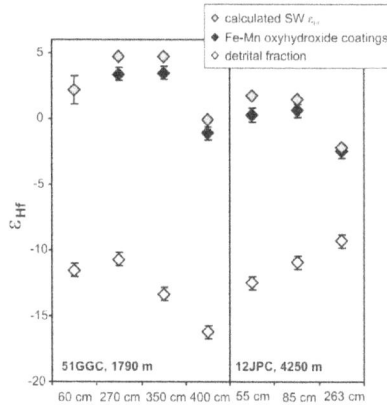

Figure 5.3. Hafnium isotope compositions of the detrital and the Fe-Mn oxyhydroxide fractions used for the mass balance calculations. The calculated hypothetical seawater ε_{Hf} is also displayed. In contrast to the mass balance calculations carried out in chapter 2 the calculated compositions are probably misleading. See text for further discussion.

The detrital fraction in the sediments of the Blake Ridge range in ε_{Hf} from -16.1 to -9.2, significantly less radiogenic than the Fe-Mn oxyhydroxide fractions. The agreement between the measured and calculated values is very good (Fig. 5.3), particularly given the large isotopic contrast but small difference in Hf concentration between the Fe-Mn oxyhydroxide fractions and the detritus. In this respect this approach would not be expected to yield the same level of consistency found for Nd and Pb. There is indeed a small offset between the measured Hf isotope compositions in the Fe-Mn oxyhydroxide fractions and the calculated seawater Hf isotope compositions which ranges from 0.03 ε_{Hf} to 1.43 ε_{Hf}. Unlike with the calculated

Table 5.3. *Measured Sr and Hf concentrations, isotope ratios, and individual results of the mass balance calculations*

Sample #	$^{87}Sr/^{86}Sr$	Sr [ng/g]	Hf [ng/g]	f (Eq. 1)	$Sr_{detr, coating}$ [ng/g] (Eq. 2)	p (Eq. 3)	$Hf_{detr, total}$ [ng/g] (Eq. 4)	$Hf_{detr, coating}$ [ng/g] (Eq. 5)	$Hf_{SW, coating}$ [ng/g] (Eq. 6)	measured ε_{Hf} (detritus)	measured ε_{Hf} (mix)	calculated ε_{Hf} (seawater) (Eq. 8)	Δ sw - mix
Core 51GGC-													
60 cm	0.70918	14059	24	0.00011	2	0.003%	2001	0.1	24	-11.40	2.18	2.21	0.03
60 cm detritus	0.72076	58259	2001										
270 cm	0.70958	2451	38	0.03019	74	0.085%	3874	3.3	35	-10.60	3.41	4.74	1.34
270 cm detritus	0.72228	86659	3874										
350 cm	0.70977	2181	41	0.03404	74	0.075%	3747	2.8	38	-13.21	3.47	4.70	1.23
350 cm detritus	0.72659	98689	3747										
400 cm	0.70983	1130	48	0.04715	53	0.054%	5718	3.1	45	-16.08	-1.10	-0.06	1.04
400 cm detritus	0.72307	98200	5718										
Core 121PC-													
55 cm	0.71500	1368	91	0.29973	410	0.301%	3069	9.2	82	-12.38	0.30	1.72	1.43
55 cm detritus	0.72861	135697	3069										
85 cm	0.71391	1121	91	0.22109	248	0.159%	4072	6.5	85	-10.85	0.63	1.50	0.87
85 cm detritus	0.73057	155606	4072										
263 cm	0.71239	870	100	0.14025	122	0.083%	4780	4.0	96	-9.19	-2.48	-2.20	0.28
263 cm detritus	0.73207	146781	4780										

seawater Nd and Pb isotope compositions, the mismatch between the measured ε_{Hf} of the Fe-Mn oxyhydroxide coating fractions and the calculated seawater Hf isotope composition is bigger than the external reproducibility for several samples shown in Table 5.3. Therefore, the uncertainties on the compositions need to be treated with more caution.

5.3.4   Seawater Hf isotope trends since the LGM

The Hf isotope evolution, as extracted by reductive leaching from Fe-Mn oxyhydroxide coatings in marine sediments along the Blake Ridge, shows a general trend from unradiogenic compositions during the LGM to more radiogenic compositions in the Holocene. In Figure 5.4a, the Hf isotope evolution is shown for core 51GGC (1790 m water depth). Lowest ε_{Hf} values of -1.7 were measured prior to the LGM. The Hf isotope compositions became significantly more radiogenic after the LGM and the section spanning the Deglaciation recorded the most radiogenic ε_{Hf} with mean values of 3.9 to 5.6 between 15.3 to 11.1 ka BP. However, this interval in the sediment core represents a section with poor reproducibility of the data. Individual ε_{Hf} values of repeated leaches of the Fe-Mn oxyhydroxide coatings of the same samples differ by between 2.4 and 3.5 ε_{Hf} (Fig. 5.4, Table 5.1). The possible reason for the poor reproducibility is discussed below. In two steps, Hf then became less radiogenic. The data reproduced better during the Younger Dryas and the transition to the Holocene. From a low in ε_{Hf} of 0.9 at 10.3 ka BP compositions then have risen towards a present-day ε_{Hf} of 2.6 ± 0.6.

Figure 5.4b illustrates the Hf isotope evolution in core 12JPC from 4250 m water depth spanning the past 20 kyr. The general pattern is similar to the Hf isotope record from 51GGC. Hafnium isotope compositions were lowest during the LGM, follow a rapid increase towards radiogenic values and level out at radiogenic ε_{Hf} values today. The least radiogenic ε_{Hf} was recorded late in the LGM at 18.7 ka BP (ε_{Hf} = -3.1). Within 3 kyr Hf isotope compositions rose to values as high as ε_{Hf} of 1.8. The deglacial Hf isotope record is isotopically variable, diplaying a total range of 2.1 ε_{Hf}. The Younger Dryas recorded ε_{Hf} as low as -0.2. Early Holocene Hf isotope compositions rose towards a present-day ε_{Hf} of 1.5. In contrast to several sampled

132

sections in shallow core 51GGC, the duplicate Hf data from Fe-Mn oxyhydroxide samples reproduced remarkably well (Fig. 5.4b).

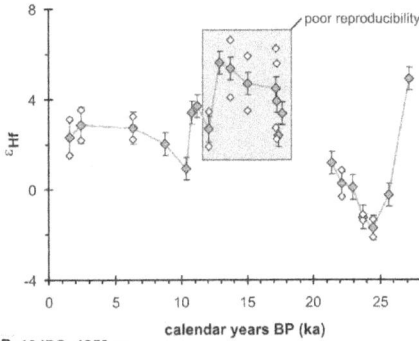

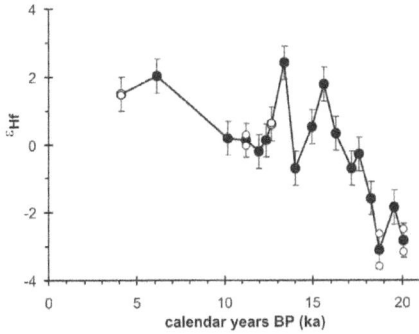

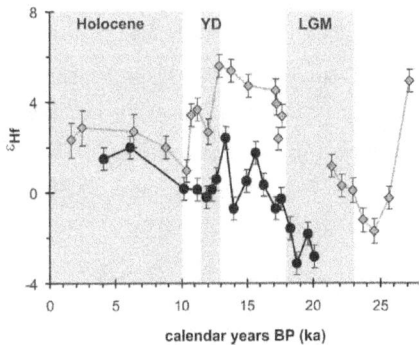

Figure 5.4. Hafnium isotope evolution for cores (a) 51GGC in 1790 m water depth and (b) 12JPC in 4250 m water depth as extracted from Fe-Mn oxyhydroxide coatings. In levels for which duplicates were obtained the individual results are plotted as white symbols in (a) and (b). Mean values were taken to illustrate the Hf isotopic trends. (c) Hafnium isotope evolution of both cores plotted in one diagram for illustration of the temporal relationships between the Hf isotope variations seen in the different depths.

In Figure 5.4c, the Hf isotope records extracted from the Fe-Mn oxyhydroxide coatings are compared. Although the two records follow approximately the same trends, the shift from least to most radiogenic compositions occurred earlier in the shallow core. The Hf isotope data converged towards an ε_{Hf} of 2 ± 1 during the Holocene, whereby core 12JPC displays slightly lower ε_{Hf} than shallow core 51GGC. A Hf isotope excursion towards less radiogenic ε_{Hf} during the Younger Dryas can be observed in both the deep and the shallow core. Overall, most Hf isotope compositions recorded at different depths in the two cores agree with reported Hf isotope compositions from ferromanganese crusts, which range from approximately -0.9 in crust ALV539 at about 250 ka (Lee et al., 1998) to ε_{Hf} of 4.1 in crust BM1969 at 2.9 Ma (Piotrowski et al., 2000; van de Flierdt et al., 2002). The sections of core 51GGC preceding the LGM and the Deglaciation appear anomalously high and might indeed represent leaching artefacts, which is also reflected by the relatively bad reproducibility of the deglaciation data in core 51GGC.

5.3.5   Pb-Nd-Hf isotope trends since the LGM

The comparison between the Hf isotope evolution seen in cores 51GGC and 12JPC on the one hand and the corresponding seawater Pb and Nd isotope trends presented earlier in chapters 3 and 4 helps determining in how far the different isotope systems are coupled. The $^{206}Pb/^{204}Pb$, ε_{Nd} and ε_{Hf} isotope evolutions were extracted from the same Fe-Mn oxyhydroxide coating aliquots and are displayed in Figure 5.5a for core 12JPC and in Figure 5.5b for core 51GGC. Note that the Nd isotope compositions are plotted with its scale reversed from that of Hf. Compared with the seawater $^{206}Pb/^{204}Pb$ isotope trends, the Hf isotope composition in Figure 5.5 only correlates to the extent that both isotope systems change from unradiogenic to radiogenic compositions in both cores in the transition to the Holocene. The timing of changes in Pb and Hf isotope compositions, as well as the short-term trends seen in the Pb and Hf isotope records, are clearly complex and often offset from each other. The changes towards more radiogenic ε_{Hf} systematically precede those observed for $^{206}Pb/^{204}Pb$ in both cores.

The Nd and Hf isotope compositions change in an inverse sense (Figure 5.5a and 5.5b). This behaviour of the isotope systems between ε_{Nd} and ε_{Hf} was very pronounced at the depth of core 51GGC (Fig. 5.5b) and is somewhat less obvious for core 12JPC (Fig. 5.5a).

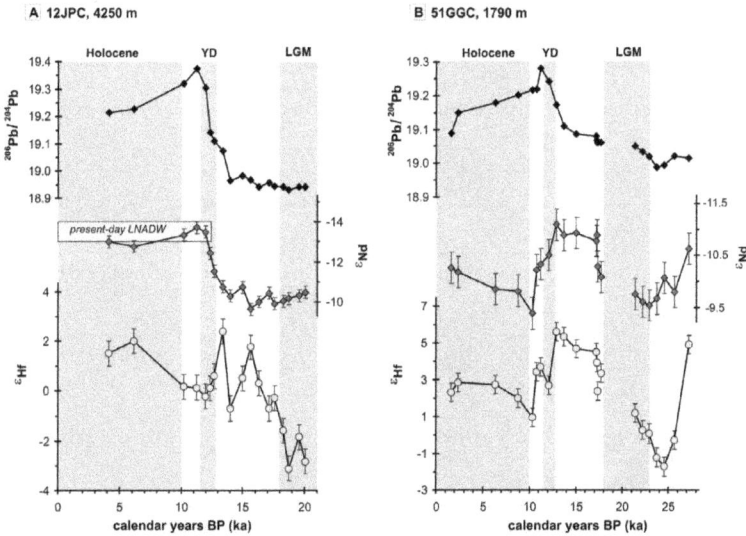

Figure 5.5. Combined $^{206}Pb/^{204}Pb$, ε_{Nd} and ε_{Hf} isotope trends derived from the same aliquots of extracted Fe-Mn oxyhydroxide fractions for cores (a) 12JPC and (b) 51GGC. Note that the Nd isotope compositions are plotted in reversed order.

5.3.6   Pb-Nd-Hf isotope trends in core 31GGC, 3410 m

Sediment core 31GGC, located at an intermediate depth between 51GGC and 12JPC, displays somewhat different Hf isotope trends compared with the deep and shallow sites. The age control of core 31GGC relatively weak because no radiocarbon dates exist and the chronology was only derived from oxygen isotope stratigraphy of the planktonic foraminifer *G. ruber* (Fig. 5.6). As with the other cores, the extracted Fe-Mn oxyhydroxide ε_{Hf} signal shows a pronounced variability, and the ε_{Hf} values are consistently more radiogenic than in the other two cores (Fig. 5.6, Table 5.1). The lowermost section (420 cm) yields an ε_{Hf} of 3. If this depth indeed

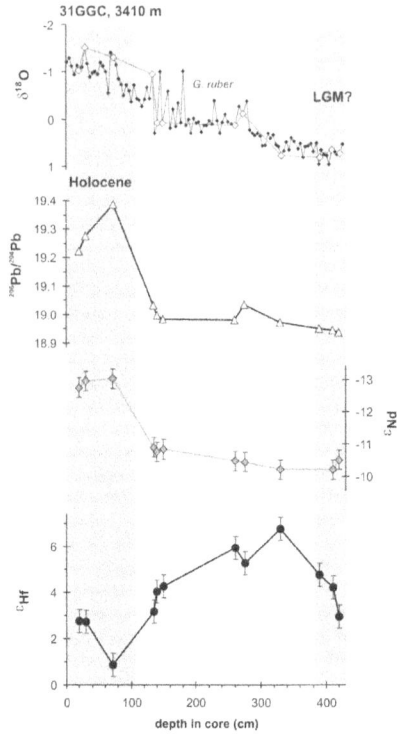

Figure 5.6. Combined oxygen stratigraphy, $^{206}Pb/^{204}Pb$, ε_{Nd} and ε_{Hf} isotope trends for core 31GGC in 3410 m water depth. The Hf isotope compositions below the Holocene sections appear high and might reflect partial leaching of the detrital phase. Note that the Nd isotope composition is plotted in reversed order. It is possible that the LGM section was not sampled in this core.

represents the LGM section, then the Hf isotope composition differs by 6.1 ε_{Hf} units from LGM compositions obtained from core 12JPC. Contrary to this large difference observed for the Hf isotope compositions, Nd isotope compositions of the same sections in 12JPC and 31GGC have identical ε_{Nd} of -10.5 and -10.2, respectively (see chapter 3). Extracted Fe-Mn oxyhydroxide Hf isotope compositions are more radiogenic in the sediments above, having ε_{Hf} as high as 6.8 in 330 cm depth. Compositions level out towards slightly lower ε_{Hf} until 135 cm depth in core and, apart from one unradiogenic composition measured for an early Holocene section in 72 cm depth in core (ε_{Hf} = 0.9), ε_{Hf} values remain moderately radiogenic around 2.9 ±

0.5. Within error the Holocene Hf isotope compositions in core 31GGC are therefore the same as those obtained from cores 12JPC and 51GGC. The inverse correlation between ε_{Hf} and ε_{Nd} seen in cores 12JPC and 51GGC is not observable in core 31GGC. The highly radiogenic compositions in the interval spanning the deglaciation are higher than any reported ferromanganese crust compositions from the North Atlantic and interpretations based on these Hf isotope compositions should thus be considered with utmost caution.

5.3.7   Neodymium-hafnium trends

Albarède et al. (1998) postulated a coupled behaviour of Hf and Nd isotopes for hydrogenetic ferromanganese crusts and by inference for seawater, which is different from crustal and mantle rocks in displaying more radiogenic ε_{Hf} for a given ε_{Nd}. The hydrogenous trend defined as the seawater array was confirmed in subsequent studies (Piotrowski et al., 2000; David et al., 2001; van de Flierdt et al., 2002). Figure 5.7 illustrates that this deviation from the terrestrial Nd and Hf isotope array is also observable for the extracted Fe-Mn oxyhydroxide fractions from the Blake Ridge sediments. From Figures 5.7b-d a relatively good agreement with compositions of ferromanganese crusts ALV539 and BM1969.05 from the western North Atlantic spanning the last 3 Myr is observed but there are also considerable differences. The Fe-Mn oxyhydroxide fractions from the three different sediment cores define slightly different trends. For core 51GGC, the inverse correlation between ε_{Hf} and ε_{Nd} described earlier (Figure 5.5b) is evident (Fig. 5.7b). In fact, the isotopic variability displayed by the Fe-Mn oxyhydroxide coating data defines a trend perpendicular to the seawater array. In contrast, the data of core 31GGC approximately follow the seawater array of Albarède et al. (1998) and overlap with the data of the Atlantic ferromanganese crusts (van de Flierdt et al., 2002), in particular for the Holocene part (Fig. 5.7c). According to the Nd isotope compositions the Holocene Fe-Mn oxyhydroxide coatings grew in NADW (ε_{Nd} = -12.9 ± 0.3), and the respective ε_{Hf} range from 0.9 to 2.8 (Table 5.1), supplying an estimate for the variability of the Hf isotope composition of Holocene NADW. Fe-Mn oxyhydroxide coatings in core 12JPC from 4250 m water depth show a similar agreement between Holocene NADW compositions and ferromanganese crust data (Fig. 5.7d). The Nd isotope composition closest to present day NADW recorded in 12JPC, represented by the least radiogenic

137

Nd isotope composition (ε_{Nd} = -13.5 ± 0.4), suggests an ε_{Hf} of 0.0 ± 0.2 of NADW in the transition to the Holocene. Extracted Fe-Mn oxyhydroxide coatings spanning the LGM and the Deglaciation follow trends similar to the data of core 51GGC (Figs. 5.7b and 5.7d). The LGM sections of cores 51GGC and 12JPC show least radiogenic ε_{Hf} corresponding to most radiogenic ε_{Nd} (Figures 5.7b and 5.7d).

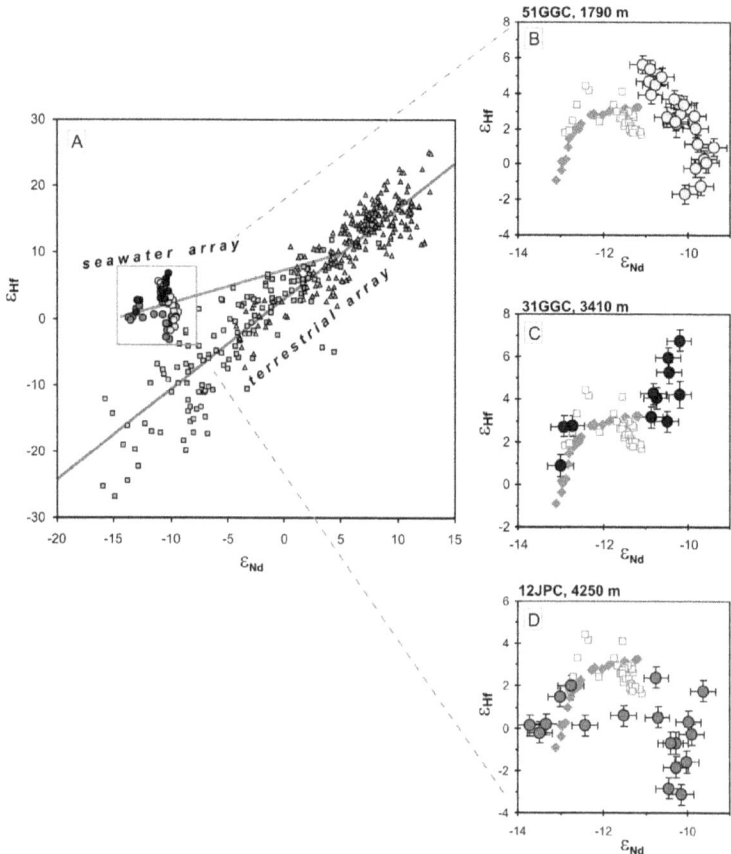

Figure 5.7. Hafnium and neodymium systematics in extracted Fe-Mn oxyhydroxide fractions from the Blake Ridge. All samples fall on the seawater array postulated by Albarède et al. (1998) for ferromanganese crust compositions. Samples for the terrestrial array are compiled in David et al. (2001). (b), (c), and (d) show blow-ups for the individual Fe-Mn oxyhydroxide fraction compositions in the three sediment cores. Also plotted are the ferromanganese crust compositions of ALV539 (dark grey diamonds) (Lee et al., 1998) and BM1969.05 (empty squares) (Piotrowski et al., 2000; van de Flierdt et al., 2002), spanning the past 3 Myr.

5.3.8   North Atlantic water column ε_{Hf} variability during the LGM

Hafnium isotope analyses were also carried out on LGM sections of four additional cores along the Blake Ridge in order to gain a better insight into the seawater Hf isotope composition during the LGM. The individual results of the measurements, as well as the mean ε_{Hf} for each sediment core are shown in Figure 5.8. Although for most cores the individual Hf isotope compositions of the Fe-Mn oxyhydroxide fractions reproduced very well, there is significant scatter for several samples. One sample in 1710 m water depth yielded an unrealistically high ε_{Hf} of +10.1. An investigation of the cause for this outlier revealed neither a technical, nor a blank contribution to this specific sample, for which reason it appears to represent a leaching artefact. Interestingly, the $^{87}Sr/^{86}Sr$ of this sample was in the same range than the other leached Fe-Mn oxyhydroxide coatings (Table 5.3). The putative LGM section in core 31GGC in 3410 m water depth presented in Figures 5.6 and 5.8 also plots outside the Hf isotope trend indicated by the LGM sections in the cores above and below. Hence if the measured Hf isotope composition indeed represents a seawater ε_{Hf} in core 31GGC it suggests that the LGM section was not sampled in core 31GGC because the ε_{Hf} is too radiogenic. The LGM sections below 2000 m water depth that yielded reproducible Hf isotope compositions suggest a mean deep water Hf isotope composition of $-2.5 \leq \varepsilon_{Hf} \leq 0.8$.

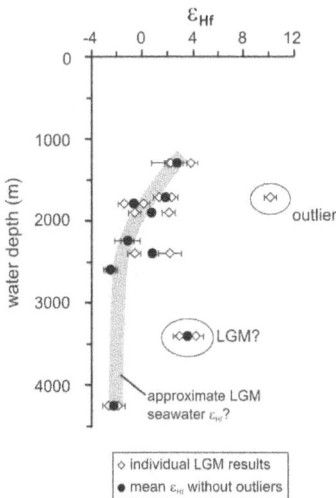

Figure 5.8. Tentative Hf isotope composition of the western North Atlantic during the LGM. Individual Hf isotope compositions of extracted Fe-Mn oxyhydroxide fractions are shown together with mean ε_{Hf} compositions neglecting one outlier in 1710 m water depth (see also Table 5.1). ε_{Hf} of core 31GGC seems to radiogenic to represent the LGM.

5.4   Discussion

Most Hf extracted through reductive leaching from the sediments along the Blake Ridge is derived from the fine fraction (clays and silts) (Fig. 5.1, section 5.3.1). This finding is not too surprising because the fine fraction in sediments offers most surface area per volume on which oxyhydroxide coatings can grow. However, the fine fraction in sediments is also the most susceptible to the release of trace metals of the detrital fraction contemporaneously with the chemical dissolution of Fe-Mn oxyhydroxide coatings from grain surfaces.   This is endorsed by the ^{87}Sr/^{86}Sr composition of the Fe-Mn oxyhydroxide leach solutions, which was often offset from seawater compositions.

In chapter 2 a number of arguments were brought forward to prove that the seawater-derived Nd, Pb and Th isotope compositions are not disturbed by this contemporaneous leaching of detrital material. It was demonstrated that due to mass balance constraints, significant contributions of the detrital fraction would be necessary to disturb the seawater isotope signal, basically because Nd, Pb and Th are highly enriched in the Fe-Mn oxyhydroxide coatings. In contrast, Hf is depleted in the Fe-Mn oxyhydroxide fractions and only low concentrations could be extracted during leaching (Table 5.2). This in itself is a positive fact because too aggressive leaching should have released more Hf from the detrital pool in the sediments, a feature that was not observed here.

There is some slightly greater level of ambiguity regarding whether the leachate compositions always define that of seawater. The Fe-Mn oxyhydroxide coating signal is compositionally and isotopically very different from the detrital fraction (Figs. 5.2 and 5.3). The purity of the seawater Hf isotope signal in some of the samples may have been compromised but most analysed samples show systematic trends yielding realistic ε_{Hf} comparable to ferromanganese crust Hf isotope compositions in the North Atlantic, and many samples reproduced well.

5.4.1   Applicability of the mass balance calculations

Although the mass balance calculations are a realistic approach for the determination of possible contributions from the detrital phase to the seawater-derived Fe-Mn oxyhydroxide fraction for Nd and Pb, it is argued here that such calculations cannot be applied in the case of Hf because of the specific chemical behaviour and the widely differing distribution of Lu and Hf in the major rock-forming minerals. The vast majority of Hf in the continental crust is stored in zircons, which consequently contain large quantities of non-radiogenic Hf (~ 1 wt%), which paired with low abundances of Lu results in a very unradiogenic Hf isotope composition. Other crustal minerals will preferentially incorporate Lu, thus creating a radiogenic Hf pool over time (Patchett et al., 1984; White et al., 1986). Zircons are known to be very weathering-resistant, and a close look at Figure 5.3 illustrates the "zircon-effect" supplying unradiogenic Hf in the detrital phase to sediments along the Blake Ridge. Fine-grained clayish marine sediments display radiogenic Hf isotope compositions (Patchett et al., 1984), and are much more susceptible to chemical partial dissolution. Therefore, should the detrital phase in sediments be attacked during the extraction of the Fe-Mn oxyhydroxide phase, then detrital Hf is most likely released from radiogenic clays and not from weathering-resistant zircons. The mathematical approach behind the mass balance calculations however seeks to quantify detrital contributions from the unradiogenic zircon-dominated detrital bulk signal. For this reason the preferential release of radiogenic Hf from a radiogenic pool *only* is not properly accounted for.

It could be argued that the $^{87}Sr/^{86}Sr$ composition of many Fe-Mn oxyhydroxide coatings deviated from the present-day seawater ratio, implying that this offset in $^{87}Sr/^{86}Sr$ was accompanied by detrital contamination of the seawater-derived Hf isotope signal. This argument can be ruled out through another look at Figure 5.1: Even for the Hf isotope compositions presented in Figure 5.1 the $^{87}Sr/^{86}Sr$ did not reliably record any significant detrital contamination. Within error, the two sediment samples chosen from the LGM section in core 51GGC (i.e., 395 cm and 412 cm) have identical ε_{Hf} for the bulk and the fine fraction, despite highly variable $^{87}Sr/^{86}Sr$ in the same aliquots. A similar Hf and Sr isotopic relationship can be observed for many duplicate samples listed in Table 5.1. Therefore the Sr isotope signal in extracted Fe-Mn oxyhydroxide coatings also does not seem to reliably reflect partial leaching of the detrital phase for Hf.

No unambiguous proof for a pure hydrogenetic origin of the Hf in the Fe-Mn oxyhydroxide fractions can be offered but because of (a) the good reproducibility of many Hf samples and (b) the good agreement with ferromanganese crust Al/Hf and Hf isotope compositions the vast majority of chemically extracted Fe-Mn oxyhydroxide Hf isotope compositions shown here indeed appear to be purely seawater-borne and to reflect the Hf isotope evolution of seawater from the LGM to present-day. In lack of further tools for the quantification of a pure seawater signal I suggest that samples with rather realistic ε_{Hf} and good reproducibility most likely reflect past seawater compositions, whereas those with bad reproducibility and very high Hf isotope compositions very likely represent leaching artefacts.

5.4.2   Seawater ε_{Hf} trends since the LGM

Applying the selection criteria defined in section 5.4.2 suggests that in core 51GGC the LGM and the Holocene sections reflect seawater ε_{Hf}. In fact, core 12JPC seems to truly reflect the seawater Hf isotope evolution throughout, since the LGM until present-day. In the following the implications of these seawater Hf isotope records will be highlighted. Hafnium isotope compositions in cores 51GGC and 12JPC in Figure 5.4 evolve from unradiogenic ε_{Hf} during the LGM to fairly radiogenic ε_{Hf} today.

On the basis of a an examination of North Atlantic ferromanganese crust Nd and Hf isotope trends over the past 3 Myr van de Flierdt et al. (2002) concluded that the Pleistocene seawater Hf isotope evolution has been largely controlled by efficient physical erosion on the continents during glacial weathering. Over the course of the Pleistocene, ε_{Hf} became less radiogenic and the youngest reported time-integrated ε_{Hf} reported from ferromanganese crust ALV539 is -0.9 (Lee et al., 1998). The lowest reported ε_{Hf} for Fe-Mn oxyhydroxide coatings from sediments along the Blake Ridge during the LGM is -1.7 (51GGC in 1790 m) and -3.1 (12JPC in 4250 m) (Table 5.1). In view of the fact that these unradiogenic ε_{Hf} values occurred during the LGM this finding strongly supports the earlier proposed zircon-grinding effect (van de Flierdt et al., 2002).

142

The suggestion of Hf release from glacially-eroded zircons is controversial. Even if rock substrate is pulverised by sub-glacial erosion underneath several kilometres of ice this does not necessarily imply that the rock flour is capable of releasing its zircon-bound Hf. The hydrological and chemical conditions underneath large ice sheets are not well understood. However, the occurrence of connected subglacial lakes has, for instance, been reported from East Antarctica (Wingham et al., 2006), providing (micro-) environments in which such chemical dissolution might occur. Zircons in old continental source areas such as the Canadian shield can be extremely unradiogenic in ε_{Hf}. In such cases only minute contributions of zircon-derived Hf are necessary to influence the seawater Hf isotope budget due to the very small concentrations of Hf in seawater, making the suggested zircon-weathering effect during the LGM feasible.

Lowest ε_{Hf} during the LGM were recorded in core 12JPC from 4250 m water depth and not at the shallow sites. This feature is surprising because the sites below 2200 m water depth along the Blake Ridge were bathed in Southern Source Water (SSW), as clearly indicated by the Nd isotope record (chapter 3) with only minor admixture of waters of northern origin. The upper water column, however, was ventilated by Glacial North Atlantic Intermediate Water (GNAIW) (Keigwin, 2004; Curry and Oppo, 2005). The most significant changes in glaciation over the past 3 Myr occurred in the northern hemisphere (Dyke et al., 2002) and the strongest negative Hf isotopic excursions should thus be detectable in waters of northern origin (i.e., GNAIW) and not in waters advecting from the Southern Ocean (i.e., SSW). This indicates that the admixture of northern source waters to the ambient SSW along the deeper the Blake Ridge, which obviously only had a minor effect on the Nd isotope composition (see chapter 3), dominated the Hf budget along the deeper Blake Ridge. In other words: The Hf isotope composition in SSW along the Blake Ridge appears to be dominated by admixture of Hf from the northern hemisphere. The question remains whether the Hf isotope signal could have been advected from the Artic Ocean, or if it was derived from more proximal sources such as Greenland or eastern North America.

Early during the deglaciation Hf isotope compositions became much more radiogenic. For example, the Hf isotope composition in core 12JPC changes within 3 kyr from ε_{Hf}

of -3.1 to 1.8. Because water mass stratification did not change during this interval and SSW dominated the deeper Blake Ridge throughout the deglaciation (see chapter 3) this change cannot be attributed to a different provenance signal but must be driven by changes in the weathering regime on the continents. The retreat of the Laurentide ice sheet was initiated soon after the LGM but it retained a large volume for most of the deglaciation until the early Holocene (Dyke and Prest, 1987; Licciardi et al., 1998). It may nevertheless be possible that the ice volume loss early during the deglaciation was significant enough to lower its effectiveness in terms of mechanical grinding of rock substrate, hence significantly reducing the unradiogenic Hf input into the North Atlantic. Reduction of glacially eroded Hf input to the North Atlantic would allow the original SSW Hf isotope composition to become more dominant. Seawater ε_{Hf} for ferromanganese crust surface scrapings of Southern Ocean waters are in the range of 1.7 to 2.9 (Godfrey et al., 1997; David et al., 2001). The Hf isotopic variations observed in core 12JPC during the early deglaciation seem to reflect the shift from a seawater Hf isotope signal dominated by North American Hf input towards original advected SSW ε_{Hf} compositions.

While van de Flierdt et al. (2002) argued that the seawater Hf isotope composition during the Pleistocene has been strongly controlled by intense mechanical erosion during glacial weathering ("zircon weathering effect"), Bayon et al. (2006) suggested that the preferential dissolution of accessory mineral phases such as apatite, sphene and allanite strongly control the marine records through riverine Hf input. This first glacial-interglacial Hf and Pb isotope record presented here enables us to better constrain which mineral phases have a stronger control on the North Atlantic Hf isotope budget.

Hafnium isotope compositions in the deep core 12JPC remained very unradiogenic until the end of the LGM and changed towards much more radiogenic compositions soon after. This finding provides support for the argument that indeed mechanical grinding of rock substrate supplied some zircon-derived unradiogenic Hf to the North Atlantic during the LGM. Zircons are unradiogenic in ε_{Hf}, contain high concentrations of U but low Th. Radiogenic Pb released from zircons should therefore supply low $^{208}Pb/^{206}Pb$, contemporaneous with unradiogenic Hf. Figure 5.9 illustrates that Hf

144

isotopes and the $^{208}Pb/^{206}Pb$ compositions of the Fe-Mn oxyhydroxide coatings are not coupled. If both the dissolved Hf and Pb budget transferred to the North Atlantic during the LGM and the deglaciation were controlled by release from weathering of zircon, then these two isotope systems should co-vary. The first significant imprint on the Pb isotope records from cores 51GGC and 12JPC however were found at the onset of the Younger Dryas (chapter 4, Fig. 5.5). Clearly Hf has a much longer residence time in seawater than Pb, introducing the complicating parameter of potential longer transport distances of Hf compared with Pb. It was argued in chapter 4 that the Pb isotope composition before the Younger Dryas remained fairly unradiogenic because continental runoff was not directly routed into the North Atlantic. Hence it is possible that the Pb signal was lost on the transect to the Blake Ridge.

The very unradiogenic ε_{Hf} during the LGM strongly suggest that glacial zircon weathering dominates the seawater Hf isotope budget. Lead is obviously not coupled to the changes in weathering regime early after the LGM (Fig. 5.5, 5.9), indicating that other U- and Th-rich mineral phases such as apatite, sphene and allanite or monazite control the proximal marine Pb isotope budget. The radiogenic Pb isotope spike observed after the Younger Dryas on the other hand does not coincide with a radiogenic Hf isotope pulse (Fig. 5.5a). Therefore it seems unlikely that the Hf budget in the western North Atlantic has been controlled by the release of Lu-rich phases such as apatite and sphene.

Considering that Hf isotope compositions remain at moderately radiogenic compositions throughout the Holocene, and have not mimicked the isotopic evolution seen in the different Pb isotopes, suggests that the Holocene Hf isotope composition recorded along the Blake Ridge reflect a near zircon-free bulk continental input signal. In this context it should be noted that, within error, ferromanganese crusts BM1969.05 and AL539 from the western North Atlantic prior to 3 Ma and predating intense northern hemisphere glaciation recorded identical integrated Hf isotope compositions of around +3 (Lee et al., 1998; Piotrowski et al., 2000). Although this observation might be coincidental, it supports the notion that chemical weathering during interglacials has supplied more incongruently weathered Hf to the North Atlantic.

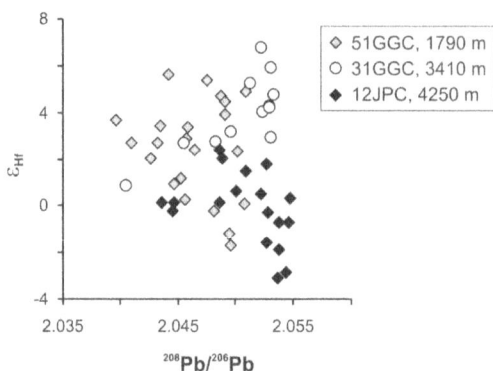

Figure 5.9. Hf isotope compositions of the Fe-Mn oxyhydroxide fractions in the three sediment cores plotted against $^{208}Pb/^{206}Pb$. If the two isotope systems were both controlled by weathering of zircons, then a co-variation should be detectable, which is not the case.

5.4.3 The "dissolved Hf" issue

The question whether the relatively high ε_{Hf} for a given ε_{Nd} in seawater compared with terrestrial rocks has in fact been caused by either significant hydrothermal Hf contributions (White et al., 1986; Godfrey et al., 1997) or the zircon effect (Piotrowski et al., 2000; van de Flierdt et al., 2002) cannot be fully clarified here. Bau and Koschinsky (2006) recently argued that truly dissolved Hf cannot be supplied to the oceans because of its speciation and particle reactivity. The presented Hf isotope compositions obtained from Fe-Mn oxyhydroxide fractions along the Blake Ridge vary over a wide range and clearly imply a dominant continental weathering source (ε_{Hf} as low as -3.1). While there is no clear evidence for a hydrothermal component in the Hf isotope signal along the Blake Ridge, it can also not be unambiguously disproved. However, because the Hf isotope composition recorded in Fe-Mn oxyhydroxide coatings along the deep Blake Ridge tightly follows climatic trends the incongruent weathering mechanism seems a much more plausible reason for the glacial-interglacial variability observed here.

Furthermore, although the Hf isotope record from the Blake Ridge is offset from the terrestrial array defined by Vervoort et al. (1999) towards the seawater array (Albarède et al., 1998), it must be strongly influenced by a continental input signal. Bau and Koschinsky (2006) argued that terrestrial Hf cannot be transferred to the oceans because of quantitative removal in the estuaries. If the Hf isotope data presented above indeed represent seawater compositions, then these conclusions should be revised. Although hydrothermal contributions may have biased the Hf isotope compositions observed along the Blake Ridge towards more radiogenic ε_{Hf}, the dominating original signal must have been continental because of its Hf isotopic range ($-3.1 \leq \varepsilon_{Hf} \leq 6.8$; the radiogenic ε_{Hf} end-member most likely reflecting partial leaching of the detrital fraction). This Hf range is clearly different from present-day MORB in the North Atlantic ($14 \leq \varepsilon_{Hf} \leq 24$) (Blichert-Toft et al., 2005) and it is unlikely that dissolution of dust has been responsible for such a large unradiogenic contribution able to balance the hydrothermal signature. Whether Hf incorporated into Fe-Mn oxyhydroxide coatings in sediments along the Blake Ridge was colloid-bound or truly dissolved on its travel path from the continental input source to the Blake Ridge cannot be decided on the basis of the data of this study. In both cases it serves as a water mass tracer.

5.5   Conclusions

This study represents the first systematic attempt to extract and characterise a seawater Hf isotope signal stored in Fe-Mn oxyhydroxide coatings of marine sediments. Although the definite proof for the seawater origin of the Hf isotope records could not be provided in this study, the consistency of the isotope records, the overall temporal and spatial trends, and the agreement with ferromanganese crust compositions suggest that seawater-derived Hf can indeed be extracted from marine sediments. Given remaining doubts, the results presently need to be interpreted with care. Given that no tool is currently available to constrain the purity of the seawater-origin of the leached Fe-Mn oxyhydroxide fractions the results presented here need confirmation through direct seawater measurements, as well as further investigations with the goal to find more reliable measures for the true seawater origin of the extracted Hf isotope signal.

Whenever results obtained from the leached Fe-Mn oxyhydroxide fraction were inconsistent and not reproducible, the general trend was of detrital contamination shifting ε_{Hf} towards more radiogenic compositions, which is most likely attributable to partial leaching of clays or detrital hematite. Most Hf isotope results obtained from

147

core 31GGC in 3410 m water depth preceding the Holocene seem affected by partial leaching of the detrital phase. The deepest core 12JPC however seems remarkably reliable and the tentative ambient seawater Hf isotope record shows the transition from glacial towards Holocene weathering patterns with yet unprecedented detail.

The tentative seawater Hf isotope compositions at the shallow part of the Blake Ridge from 1790 m water depth and the deep site at 4250 m water depth during the LGM were most likely dominated by Hf inputs from nearby northern continental sources. Epsilon Hf as low as -3.1 in deep North Atlantic waters during the LGM suggest that northern hemisphere glaciation indeed allowed for more efficient mechanical erosion during glacial weathering, also supplying a fraction of unradiogenic Hf derived from zircons to the Blake Ridge. The Hf isotope composition of the LGM section in 4250 m water depth suggests that Southern Source Water bathing the lower Blake Ridge was dominated by Hf admixtures of northern source water.

Hf isotopes are decoupled from the seawater radiogenic Pb isotope evolution recorded along the Blake Ridge during the deglaciation and the Holocene and there is no evidence that the marine Hf isotope budget in the northwest Atlantic has been controlled by preferential weathering of accessory phases such as apatite and sphene. The Hf isotope record rather seems to reflect a switch from more congruent weathering during the last glacial towards zircon-free incongruent bulk weathering during the Holocene.

The deepest core analysed along the Blake Ridge from 4250 m water depth yielded remarkably consistent and reproducible Hf isotope compositions. In 4250 m water depth Hf isotope compositions in the newly established NADW following the Younger Dryas remained at relatively unradiogenic values (ε_{Hf} of 0.0 ± 0.2) and increased in the course of the Holocene towards present-day ε_{Hf} of 1.5 ± 1. The Younger Dryas can be depicted in both 1790 m and 4250 m water depth as an excursion towards unradiogenic Hf isotope compositions.

If the Hf isotope records presented here indeed reflect seawater-borne Hf isotope trends, then the seawater Hf budget along the Blake Ridge was clearly controlled by continental input and hydrothermal contributions can only have been of minor importance.

Acknowledgements

Tina van de Flierdt compiled and calibrated Hf isotope results for crusts ALV539 and BM1969.05.

Chapter 6

General conclusions and outlook

6.1   Conclusions

The ultimate goal of this dissertation was to introduce and evaluate new paleoceanographic and paleoclimatic proxies and to directly apply these in the western North Atlantic. The archive used was the seawater-derived Fe-Mn oxyhydroxide fraction in marine pelagic sediments from the Blake Ridge spanning Marine Isotope Stages 2 and 1 with special emphasis on the transition from the Last Glacial Maximum to the Holocene. The authigenic Fe-Mn oxyhydroxide coatings incorporate trace metals from the overlying water body thereby incorporating its radiogenic (Nd, Hf, Pb, Th) isotopic composition. The advantage of this work was that different trace metals were extracted from the same sediments, which allows much more far reaching conclusions to be made compared with the stand-alone usage of any single trace metal. Seawater-derived Nd isotopes set the frame to identify the water masses prevailing in the upper and lower segment of the Blake Ridge at a given time. Once the hydrographic situation was clarified, climatic and weathering-related processes on the nearby North American continent could be analysed. The Pb isotope trends observed during and after the Younger Dryas contain valuable information about continental runoff reorganisations from the North American continent. It further highlights the slow retreat of the Laurentide ice sheet, which progressed only at moderate pace during the deglaciation and still retained a significant volume until the Younger Dryas. The glacial-interglacial seawater Hf isotope evolution provides answers to the question about which mineral phases in fact govern the Hf budget transferred to the ocean.

6.1.1   New proxy development in sedimentary archives

Earlier studies using seawater-derived Nd from Fe-Mn oxyhydroxide fractions in marine sediments used the $^{87}Sr/^{86}Sr$ isotope composition of the same fraction to prove the seawater origin of the extracted Nd. The seawater $^{87}Sr/^{86}Sr$ isotope composition however is easily disturbed during reductive chemical extraction of the Fe-Mn oxyhydroxide coatings by contributions from the detrital fraction. Contrary to $^{87}Sr/^{86}Sr$, seawater-derived Nd, Pb and Th isotope compositions are very resistant against detrital contamination because these trace metals are highly enriched in the authigenic fraction compared with the detrital pool in marine sediments.

Osmium isotopes were tested as an alternative proxy to detect detrital contamination to the Fe-Mn oxyhydroxide fraction. Similarly to $^{87}Sr/^{86}Sr$ compositions, this radiogenic isotope system produced $^{187}Os/^{186}Os$ deviating from seawater compositions. Under consideration of the very low Os concentrations extracted from marine sediments this isotope system was also found to be an unreliable indicator about the seawater origin of extracted Nd, Pb and Th.

Better constraints were found by the observation of Rare Earth Element (REE) patterns in the extracted and the residual phases in the sediments. A characteristic shale-normalised MREE enrichment was observed for the extracted Fe-Mn oxyhydroxide fractions. Although $^{87}Sr/^{86}Sr$ compositions suggested detrital contributions to the extracted seawater signal, the REE patterns were identical to marine pore water compositions under anoxic conditions, when authigenic Fe-Mn oxyhydroxide coatings dissolve releasing their trace metal load.

Additional investigations monitoring the elemental ratios of Al/Nd, Al/Pb and Al/Th in the seawater-derived Fe-Mn oxyhydroxide fraction and the detrital phase revealed the pronounced enrichment of Nd, Pb and Th in the seawater-derived fraction, and the respective elemental ratios were very similar to abyssal ferromanganese crust compositions in the Equatorial and South Pacific.

The best evidence for the robustness of the seawater Nd and Pb isotope signal towards detrital contamination was supplied by mass balance calculations. On the basis of the Sr, Nd and Pb isotope compositions in the Fe-Mn oxyhydroxide fraction and the detrital phase and the respective element concentrations in the different fractions it could be shown why the extracted Nd and Pb isotope compositions reproduced so well despite highly variable $^{87}Sr/^{86}Sr$ compositions. The mass balance calculations demonstrated that the Sr isotope composition is very easily affected by minute contributions of partial leaching of the detrital phase on the one hand, whereas Nd and Pb isotope compositions require significant contributions from the detrital pool to be leached to mask the seawater isotope composition.

Seawater-derived Nd, Pb and Th isotope compositions can be obtained from Fe-Mn oxyhydroxide coatings. Surely sediment composition is a crucial parameter with regard to reductive chemical extraction of an authigenic phase, and at different localities in the oceans the leaching behaviour might be different. The results presented in this dissertation demonstrate however that this paleoceanographic archive should be widely applicable.

6.1.2 Major North Atlantic water mass changes traced with Nd isotopes

In chapter 3 the present-day seawater Nd isotope signal along the Blake Ridge as stored in Fe-Mn oxyhydroxide coatings, the Nd isotope compositions recorded during the LGM, and downcore results at three different sites along the shallow (1790 m), intermediate (3410 m), and deeper (4250 m) Blake Ridge were presented, allowing for a tight reconstruction of the hydrographic settings during termination 1 and the transition to the Holocene.

At present-day the NADW Nd isotope composition having an ε_{Nd} of around -13.5 could be extracted from Fe-Mn oxyhydroxide coatings in sediments from within the major flow axis of the DWBC. Above the major flow axis the extracted Nd isotope signal of NADW indicated either much more radiogenic Nd isotope compositions in the water masses near the coast, or the signal has been influenced by Fe-Mn oxyhydroxide coatings of sediments carrying seawater signals from the North American shelf. A slight influence of AABW is observable at the lowermost sites in 4700 m water depth along the Blake Ridge today.

During the LGM the distinct unradiogenic NADW Nd isotope compositions could not be found at any depth along the Blake Ridge, suggesting that water advecting from the Labrador Sea did not contribute to the glacial equivalent of NADW, the Glacial North Atlantic Intermediate Water (GNAIW). The process of sediment re-distribution, which was identified applying $^{230}Th_{xs}$ in the shallow core, that offset ambient seawater Nd isotope compositions during the deglaciation and the Holocene, was not operating during the LGM. Due to this feature we interpret the extracted glacial Nd isotope compositions in the shallow core in 1790 m to truly reflect GNAIW Nd isotope compositions. The radiogenic ε_{Nd} of the GNAIW of –9.8 suggests that Nd originating from isotopically unmodified waters from either the Irminger- or the Iceland Basins in the subpolar North Atlantic fed the GNAIW along the upper Blake Ridge. The deeper part of the Blake Ridge was dominated by advected Southern Source Water (SSW); however its Nd isotopic composition is very similar to the GNAIW Nd composition dominating water depths above around 2200 m. The water column profile from the LGM supports the presence of a water mass boundary in the glacial western North Atlantic centred at around 2200 m water depth.

According to the Nd isotope records, the deeper Blake Ridge was bathed in SSW throughout the deglaciation and the modern circulation mode was initiated after the Younger Dryas.

In conjunction with the Nd isotope results from 1790 m water depth, the high sediment focusing rates during the Holocene estimated by ^{230}Th$_{xs}$ results highlight an important factor influencing the radiogenic isotope compositions incorporated into Fe-Mn oxyhydroxide fractions: The original ambient seawater composition seemingly can be altered at continental rise settings due to contributions from above, especially in drift deposits such as the Blake Ridge. This apparent surface water contribution is also observable in the Pb isotope record presented in chapter 4.

6.1.3 Constraining the timing of continental runoff reorganisations from eastern North America

In the studied time interval and this particular geographic setting, the seawater Pb isotope evolution along the Blake Ridge is only in part a function of provenance. Much more importantly it reflects climate, the dominant weathering regime on the Canadian Shield, and sensitively traces freshwater drainage during the deglaciation. Pb isotope compositions in every monitored water depth were very unradiogenic during the last glacial and remained relatively unradiogenic during most of the deglaciation. At the onset of the Younger Dryas the Pb isotope compositions both at the lower and the upper Blake Ridge changed dramatically within very short time and a short-lived very radiogenic Pb spike is observed at 11.2 ka calendar BP. Because Pb isotope composition of seawater during the deglaciation are a direct function of the freshwater input in the North Atlantic this short-lived radiogenic Pb spike was most likely caused by freshwater/meltwater diversions into the western North Atlantic. During the Holocene the Pb isotope compositions remained relatively radiogenic,

154

reflecting of the establishment of the Holocene chemical weathering regime on the Canadian Shield.

This first glacial-interglacial seawater Pb isotope record at sub-millennial resolution clearly corroborates the incongruent release of radiogenic Pb during weathering on the continents. Water advecting through the Labrador Sea prior to it arrival at the Blake Ridge carried the lowest $^{207}Pb/^{206}Pb$ and $^{208}Pb/^{206}Pb$ during the early Holocene, contemporaneous with lowest ε_{Nd}, a feature that points to incongruent preferential release of radiogenic ^{206}Pb (as well as ^{208}Pb). The factors controlling the Pb isotope compositions in the transition to the Holocene is probably (a) washout of loosely bound alpha-recoiled radiogenic Pb from damaged lattices sites of rock substrate freshly crushed during the last glacial, and (b) intensification of chemical weathering following climate-driven northward migration of the permafrost in North America leading to preferential dissolution of U- and Th-rich accessory minerals such as allanite, apatite and sphene that maintain a radiogenic Pb isotope composition in the western North Atlantic throughout the Holocene. Which of the processes in fact governed the Pb isotope excursion seen in the North Atlantic throughout and after the Younger Dryas could not be identified yet.

The very good agreement between seawater Pb isotope compositions during the LGM recorded in Fe-Mn oxyhydroxide coatings and those of ferromanganese crusts predating the onset of major northern hemisphere glaciation 2.7 Myr ago possibly yields information about temporal trends in continental Pb supply to the North Atlantic. The increased input of continent-derived Pb recorded in the western North Atlantic dominantly occurred during interglacials and interstadials, and was probably much lower during glacials.

The incorporation of a surface seawater signal in Fe-Mn oxyhydroxide coatings along the Blake Ridge mentioned earlier is also reflected in the Pb isotope composition of these coatings. Comparing $^{207}Pb/^{206}Pb$ and $^{208}Pb/^{206}Pb$ compositions in the shallow sediment core in 1790 m water depth with the respective Pb isotope compositions of crust "Blake" analysed by Reynolds et al. (1999), which grew on the Blake Plateau in Florida Current water, reveals a striking similarity. This feature also explains why the

155

most original Nd and Pb isotope compositions and trends were observed in the deepest core.

6.1.4   Glacial-interglacial Hf weathering trends

The extraction of a seawater Hf isotope signal from Fe-Mn oxyhydroxide coatings in Blake Ridge sediments was technically the most delicate task because only small amounts could be extracted through reductive leaching. While for Nd, Pb and Th a strong enrichment could be observed in the authigenic seawater-derived fraction, this was not the case for Hf. Indeed, Hf is depleted relative to Nd, Pb and Th in the Fe-Mn oxyhydroxide fraction. Due to its small concentrations, however, it should be very susceptible to contemporaneous leaching of the detrital phase in the sediment.

Considering the small Hf concentrations in the extracted Fe-Mn oxyhydroxide fractions, overall the Hf isotope results were strikingly reproducible and consistent. On the other hand some outliers were produced, which seemingly represent leaching artefacts. There are clear indications that most Hf isotope results represent seawater compositions, but because its origin cannot be proved with the used methods, the results presented in this dissertation remain tentative until confirmation through direct seawater measurements is available.

The deep and the shallow seawater Hf isotope compositions follow similar trends, from unradiogenic ε_{Hf} during the last glacial towards more radiogenic ε_{Hf} at present-day. The shallow site in 1790 m further recorded a very radiogenic excursion during the deglaciation, but because Hf isotope results were only poorly reproducible during this interval they should be interpreted with caution. The Hf isotope compositions for all Fe-Mn oxyhydroxide fractions is relatively radiogenic for a given ε_{Nd} and fall on the seawater array introduced by Albarède et al. (1998).

The glacial-interglacial Hf isotope trends and the very unradiogenic seawater Hf isotope compositions during the last glacial strongly support the zircon-grinding effect introduced by van de Flierdt et al. (2002), which leads to a more congruent release of Hf from bulk continental crust during intense glacial erosion. The trend towards more radiogenic ε_{Hf} shortly after the LGM might imply that the Hf isotope composition of seawater in the western North Atlantic reacts very sensitively to ice volume changes of the Laurentide ice sheet.

The Hf isotope data presented here yield no clear indication that accessory minerals such as apatite or sphene dominate the Hf budget along the Blake Ridge at any time. The glacial-interglacial Hf isotope trends rather indicate a switch from more congruent weathering during the LGM towards largely zircon-free incongruent bulk weathering during the Holocene. If the Hf isotope records presented here indeed reflect seawater-borne Hf isotope trends, then the seawater Hf budget along the Blake Ridge was clearly controlled by continental input and hydrothermal contributions can only have been of minor importance.

6.2   Outlook

The radiogenic isotope systems used here for paleoceanographic and paleoclimatic reconstruction in the western North Atlantic focused on the past 27 kyr. Obviously these records should be extended back over the entire glacial cycle and termination II considering the outstanding short-term fluctuations seen in the seawater Nd isotope composition in the Cape Basin of the South Atlantic (Piotrowski et al., 2005). The use of Hf and Pb isotopes extracted from the authigenic fraction in marine sediments allows a completely new insight into paleoclimatic processes which could not be detected using other proxies.

6.2.1   Specific suggestions for future research

Meridional circulation during the last glacial cycle in the North Atlantic
Is deep NADW ventilation only an interglacial feature? A striking feature of the Nd isotope records presented in chapter 3 was the absence of deeply ventilating NADW below 3400 m water depth throughout the deglaciation. This dissertation only covers the past 20 kyr of North Atlantic ventilation below 3410 m water depth along the Blake Ridge. How did the North Atlantic operate during interstadials in MIS 3 and the remainder of the last glacial cycle?

GNAIW ventilation depth along the Blake Ridge during the deglaciation
Were there water mass stratification changes during the deglaciation in intermediate depths (2000-3000 m)? For instance, no clear indication was found during this research for the collapse of the meridional overturning circulation during Heinrich event 1 in the shallow and deep cores, but maybe these depths were just not in the right water depth to resolve it. What exactly happened in water depths between 2000 and 3000 m during the deglaciation? Can the seawater Nd isotope record be brought in agreement with the radiocarbon ventilation age records from intermediate depths presented by Robinson et al. (2005) or the Cd/Ca-δ^{13}C record of Rickaby and Elderfield (2005)?

Further evidence for shutdown of Labrador Sea Water formation
The Nd isotope records from the Blake Ridge suggest that Labrador Sea Water did not contribute to GNAIW during the LGM. How long was Labrador deep water formation shut off and this fully glacial circulation mode operating? When and how was it initiated?

Further evidence for freshwater fluxes into the North Atlantic
The radiogenic Pb isotope spike seen immediately following the Younger Dryas was probably caused by a large scale meltwater diversion event. Which route did the water take from the North American continent? Was it entrained in Labrador Sea Water or added subsequently through the St. Lawrence and Hudson Rivers? How much freshwater was drained into the Arctic Ocean during the deglaciation, especially during the Younger Dryas and afterwards?

The last interglacial - similar to the Holocene?
The oxygen isotope record from NGRIP (Fig. 1.3) suggests even warmer conditions at peak interglacial conditions during MIS 5e. In how much was this period similar to the current climate? Would Nd, Hf and Pb isotope compositions follow similar trends and identical compositions; was deep NADW ventilation in operation analogously to today?

Variability of the Laurentide ice sheet volume

The Hf isotope records in the transition from the LGM to the deglaciation become more radiogenic within only few thousand years, despite the fact that the Laurentide and Innuitian ice sheets retained large volumes for a significant time throughout the deglaciation. What is the cause for this early Hf isotope excursion? If Hf isotope compositions in seawater indeed react so sensitively to ice volume change it should be used for the quantification of ice volume changes throughout the entire last glacial cycle. Which Hf isotope compositions were recorded in MIS 3 seawater?

6.2.2   General suggestions for future research

Groundtruthing of the Fe-Mn oxyhydroxide Nd and Hf isotope signals

Although the Nd and Hf isotope signals presented here displayed a remarkably systematic and consistent behaviour a calibration of this seawater archive against foraminifera, deep sea coral records and seawater samples should be carried out to detect possible traps inherent to the individual archives. With regard to the Nd isotope results presented here the biggest threat for reliable seawater reconstructions is the extraction of pre-formed Fe-Mn oxyhydroxides transported on re-distributed sediment towards the sampling site. Even though these pre-formed Fe-Mn oxyhydroxides can be authigenic, they will certainly bias the in situ formed seawater signal if these previously formed coatings initially grew in a different water mass.

For Pb isotopes, however, direct comparison with present-day seawater cannot be done because its natural Pb isotope composition is fully obliterated by anthropogenic Pb input.

Hafnium isotopes in seawater

Should the seawater-origin of the extracted Fe-Mn oxyhydroxide fractions be confirmed in subsequent studies this would make reductive leaching of marine sediments a relatively fast new method compared with direct seawater Hf isotope measurements. Using 1 to 2 gram of bulk sediment would allow the determination of the Hf isotope composition of the overlying water mass at an external reproducibility

159

of 0.3 to 0.5 ε_{Hf}, for which a comparative amount of at least 40 to 60 l of seawater are required.

Climatic and weathering-related trends in the southern hemisphere

This dissertation is a Northern Hemisphere circulation and climate record. Southern Component water advected far into the North Atlantic during the LGM, and it is not clear how variable this water mass was in terms of Nd and Hf isotope compositions on glacial-interglacial timescales. Piotrowski et al. (2005) provide first information in this context for the Nd isotope variability, but how much did the Hf isotope composition in the Southern Ocean vary between the LGM and today? How do Pb isotope seawater records around New Zealand and the Drake Passage, representing intermediate southern latitudes, vary on glacial-interglacial timescales?

Diagenetic restrictions to the usage of Fe-Mn oxyhydroxide fractions

How far back in time can Fe-Mn oxyhydroxide coatings in sediments be used for paleoceanographic purposes? When does prolonged diagenesis obliterate the original seawater signal?

References

Abouchami, W. and Goldstein, S. L., 1995. A lead isotopic study of circum-Antarctic manganese nodules. Geochimica et Cosmochimica Acta, 59 (9), 1809-1820.

Abouchami, W., Galer, S. J. G. and Koschinsky, A., 1999. Pb and Nd isotopes in NE Atlantic Fe-Mn crusts: Proxies for trace metal paleosources and paleocean circulation. Geochimica et Cosmochimica Acta, 63 (10), 1489-1505.

Albarède, F., Simonetti, A., Vervoort, J. D., Blichert-Toft, J. and Abouchami, W., 1998. A Hf-Nd isotopic correlation in ferromanganese nodules. Geophysical Research Letters, 25 (20), 3895-3898.

Alleman, L. Y., Veron, A. J., Church, T. M., Flegal, A. R. and Hamelin, B., 1999. Invasion of the abyssal North Atlantic by modern anthropogenic lead. Geophysical Research Letters, 26(10), 1477-1480.

Alley, R. B., 2000. The Younger Dryas cold interval as viewed from central Greenland. Quaternary Science Reviews, 19 (1-5), 213-226.

Andersen, K. K., Azuma, N., Barnola, J. M., Bigler, M., Biscaye, P., Caillon, N., Chappellaz, J., Clausen, H. B., DahlJensen, D., Fischer, H., Fluckiger, J., Fritzsche, D., Fujii, Y., Goto-Azuma, K., Gronvold, K., Gundestrup, N. S., Hansson, M., Huber, C., Hvidberg, C. S., Johnsen, S. J., Jonsell, U., Jouzel, J., Kipfstuhl, S., Landais, A., Leuenberger, M., Lorrain, R., Masson-Delmotte, V., Miller, H., Motoyama, H., Narita, H., Popp, T., Rasmussen, S. O., Raynaud, D., Rothlisberger, R., Ruth, U., Samyn, D., Schwander, J., Shoji, H., Siggard-Andersen, M. L., Steffensen, J. P., Stocker, T., Sveinbjornsdottir, A. E., Svensson, A., Takata, M., Tison, J. L., Thorsteinsson, T., Watanabe, O., Wilhelms, F. and White, J. W. C., 2004a. High-resolution record of Northern Hemisphere climate extending into the last interglacial period. Nature, 431 (7005), 147-151.

Andersen, M. B., Stirling, C. H., Potter, E. K. and Halliday, A. N., 2004b. Toward epsilon levels of measurement precision on U-234/U-238 by using MC-ICPMS. International Journal of Mass Spectrometry, 237 (2-3), 107-118.

Anderson, R. F., Bacon, M. P. and Brewer, P. G., 1983a. Removal of ^{230}Th and ^{231}Pa at ocean margins. Earth and Planetary Science Letters, 66, 73-90.

Anderson, R. F., Bacon, M. P. and Brewer, P. G., 1983b. Removal of ^{230}Th and ^{231}Pa from the open ocean. Earth and Planetary Science Letters, 62 (1), 7-23.

Andersson, P. S., Dahlqvist, R., Ingri, J. and Gustafsson, O., 2001. The isotopic composition of Nd in a boreal river: A reflection of selective weathering and colloidal transport. Geochimica et Cosmochimica Acta, 65 (4), 521-527.

Anselmetti, F. S., Eberli, G. P. and Ding, Z. D., 2000. From the Great Bahama Bank into the Straits of Florida: A margin architecture controlled by sea-level fluctuations and ocean currents. Geological Society of America Bulletin, 112 (6), 829-844.

Bard, E., Hamelin, B., Arnold, M., Montaggioni, L., Cabioch, G., Faure, G. and Rougerie, F., 1996. Deglacial sea-level record from Tahiti corals and the timing of global meltwater discharge. Nature, 382 (6588), 241-244.

Bau, M., 1996. Controls on the fractionation of isovalent trace elements in magmatic and aqueous systems: Evidence from Y/Ho, Zr/Hf, and lanthanide tetrad effect. Contributions to Mineralogy and Petrology, 123 (3), 323-333.

Bau, M. and Koschinsky, A., 2006. Hafnium and neodymium isotopes in seawater and in ferromanganese crusts: The "element perspective". Earth and Planetary Science Letters, 241 (3-4), 952-961.

Bayon, G., German, C. R., Boella, R. M., Milton, J. A., Taylor, R. N. and Nesbitt, R. W., 2002. An improved method for extracting marine sediment fractions and its application to Sr and Nd isotopic analysis. Chemical Geology, 187 (3-4), 179-199.

Bayon, G., German, C. R., Burton, K. W., Nesbitt, R. W. and Rogers, N., 2004. Sedimentary Fe-Mn oxyhydroxides as paleoceanographic archives and the role of aeolian flux in regulating oceanic dissolved REE. Earth and Planetary Science Letters, 224 (3-4), 477-492.

Bayon, G., Vigier, N., Burton, K. W., Brenot, A., Carignan, J., Etoubleau, J. and Chu, N.-C., 2006. The control of weathering processes on riverine and seawater hafnium isotope ratios. Geology, 34 (6), 433-436, doi: 10.1130/G22130.1.

Belshaw, N. S., Freedman, P. A., O'Nions, R. K., Frank, M. and Guo, Y., 1998. A new variable dispersion double-focusing plasma mass spectrometer with performance illustrated for Pb isotopes. International Journal of Mass Spectrometry, 181, 51-58.

Birck, J. L., Barman, M. R. and Capmas, F., 1997. Re-Os isotopic measurements at the femtomole level in natural samples. Geostandards Newsletter, 21 (1), 19-27.

Biscaye, P. E., Anderson, R. F. and Deck, B. L., 1988. Fluxes of particles and constituents to the eastern United States continental slope and rise - Seep-I. Continental Shelf Research, 8 (5-7), 855-904.

Biscaye, P. E., Flagg, C. N. and Falkowski, P. G., 1994. The shelf edge exchange processes experiment, Seep-II - an introduction to hypotheses, results and conclusions. Deep-Sea Research Part II-Topical Studies in Oceanography, 41 (2-3), 231-252.

Blichert-Toft, J. and Albarède, F., 1997. The Lu-Hf isotope geochemistry of chondrites and the evolution of the mantle-crust system. Earth and Planetary Science Letters, 148, 243-258.

Blichert-Toft, J., Agranier, A., Andres, M., Kingsley, R., Schilling, J. G. and Albarede, F., 2005. Geochemical segmentation of the Mid-Atlantic Ridge north of Iceland and ridge-hot spot interaction in the North Atlantic. Geochemistry Geophysics Geosystems, 6.

Boyle, E. A. and Keigwin, L. D., 1987. North-Atlantic thermohaline circulation during the past 20,000 years linked to high-latitude surface-temperature. Nature, 330 (6143), 35-40.

Boyle, E. A., 1988. Cadmium: Chemical tracer of deepwater paleoceanography. Paleoceanography, 3, 471-489.

Boyle, E. A., 1992. Cadmium and delta-C-13 paleochemical cean distributions during the stage-2 glacial maximum. Annual Review of Earth and Planetary Sciences, 20, 245-287.

Broecker, W. S. and Peng, T.-H., 1982. Tracers in the Sea. Eldigio Press, Palisades NY.

Broecker, W. S., Peteet, D. M. and Rind, D., 1985. Does the ocean-atmosphere System have more than one stable mode of operation. Nature, 315 (6014), 21-26.

Broecker, W. S., Andree, M., Bonani, G., Wolfli, W., Oeschger, H., Klas, M., Mix, A. C. and Curry, W., 1988a. Preliminary estimates for the radiocarbon age of deep water in the glacial ocean. Paleoceanography, 3, 659-669.

Broecker, W. S., Andree, M., Wolfli, W., Oeschger, H., Bonani, G., Kennett, J. and Peteet, D., 1988b. The chronology of the last deglaciation: Implications to the cause of the Younger Dryas event. Paleoceanography, 3, 1-19.

Broecker, W. S. and Denton, G. H., 1989. The role of ocean-atmosphere reorganizations in glacial cycles. Geochimica et Cosmochimica Acta, 53 (10), 2465-2501.

Broecker, W. S., Kennett, J. P., Flower, B. P., Teller, J. T., Trumbore, S., Bonani, G. and Wolfli, W., 1989a. Routing of meltwater from the Laurentide Ice Sheet during the Younger Dryas cold episode. Nature, 341 (6240), 318-321.

Broecker, W. S., 1991. The great ocean conveyor. Oceanography, 4, 79-89.

Broecker, W. S., 2006. Was the Younger Dryas triggered by a flood? Science, 312 (5777), 1146-1148.

Burton, K. W., Ling, H.-F. and O'Nions, R. K., 1997. Closure of the Central American Isthmus and its effect on deep-water formation in the North Atlantic. Nature, 386, 382-385.

Burton, K. W., 2006. Global weathering variations inferred from marine radiogenic isotope records. Journal of Geochemical Exploration, Extended Abstracts presented at the 7th Symp. on the Geochemistry of the Earth's Surface (GES-7), 88(1-3), 262-265.

Chappell, J. and Shackleton, N. J., 1986. Oxygen isotopes and sea-level. Nature, 324 (6093), 137-140.

Charles, C. D. and Fairbanks, R. G., 1992. Evidence from Southern Ocean sediments for the effect of North Atlantic deep-water flux on climate. Nature, 355 (6359), 416-419.

Chen, F. K., Siebel, W. and Satir, M., 2002. Zircon U-Pb and Pb-isotope fractionation during stepwise HF acid leaching and geochronological implications. Chemical Geology, 191 (1-3), 155-164.

Chester, R. and Hughes, M. J., 1967. A chemical technique for the separation of ferromanganese minerals, carbonate minerals and adsorbed trace elements for pelagic sediments. Chemical Geology, 2, 249-262.

Chow, T. J. and Patterson, C. C., 1962. The occurrence and significance of lead isotopes in pelagic sediments. Geochimica et Cosmochimica Acta, 26 (2), 263-308.

Christensen, J. N., Halliday, A. N., Godfrey, L. V., Hein, J. R. and Rea, D. K., 1997. Climate and ocean dynamics and the lead isotopic records in Pacific ferromanganese crusts. Science, 277, 913-918.

Clark, P. U., Alley, R. B., Keigwin, L. D., Licciardi, J. M., Johnsen, S. J. and Wang, H. X., 1996. Origin of the first global meltwater pulse following the last glacial maximum. Paleoceanography, 11 (5), 563-577.

Clark, P. U., Marshall, S. J., Clarke, G. K. C., Hostetler, S. W., Licciardi, J. M. and Teller, J. T., 2001. Freshwater forcing of abrupt climate change during the last glaciation. Science, 293 (5528), 283-287.

Clark, P. U., Mitrovica, J. X., Milne, G. A. and Tamisiea, M. E., 2002. Sea-level fingerprinting as a direct test for the source of global meltwater pulse IA. Science, 295 (5564), 2438-2441.

Cochran, J. K., McKibbinvaughan, T., Dornblaser, M. M., Hirschberg, D., Livingston, H. D. and Buesseler, K. O., 1990. Pb-210 Scavenging in the North-Atlantic and North Pacific Oceans. Earth and Planetary Science Letters, 97 (3-4), 332-352.

Cohen, A. S., Onions, R. K., Siegenthaler, R. and Griffin, W. L., 1988. Chronology of the pressure-temperature history recorded by a granulite terrain. Contributions to Mineralogy and Petrology, 98 (3), 303-311.

Cottet-Puinel, M., Weaver, A. J., Hillaire-Marcel, C., de Vernal, A., Clark, P. U. and Eby, M., 2004. Variation of Labrador Sea Water formation over the Last Glacial cycle in a climate model of intermediate complexity. Quaternary Science Reviews, 23 (3-4), 449-465.

Craig, H., Krishnas.S and Somayaju.Bl, 1973. Pb-210 - Ra-226 - Radioactive disequilibrium in deep-sea. Earth and Planetary Science Letters, 17 (2), 295-305.

Creaser, R. A., Papanastassiou, D. A. and Wasserburg, G. J., 1991. Negative thermal ion mass-spectrometry of osmium, rhenium, and iridium. Geochimica et Cosmochimica Acta, 55 (1), 397-401.

Curry, W. B. and Oppo, D. W., 2005. Glacial water mass geometry and the distribution of delta C-13 of Sigma CO_2 in the western Atlantic Ocean. Paleoceanography, 20 (1), PA1017, doi:10.1029/2004PA001021.

Dahlqvist, R., Andersson, P. S. and Ingri, J., 2005. The concentration and isotopic composition of diffusible Nd in fresh and marine waters. Earth and Planetary Science Letters, 233 (1-2), 9-16.

Dansgaard, W., Johnsen, S. J., Clausen, H. B., Dahljensen, D., Gundestrup, N. S., Hammer, C. U., Hvidberg, C. S., Steffensen, J. P., Sveinbjornsdottir, A. E., Jouzel, J. and Bond, G., 1993. Evidence for general instability of past climate from a 250-kyr ice-core record. Nature, 364 (6434), 218-220.

David, K., Frank, M., O'Nions, R. K., Belshaw, N. S. and Arden, J. W., 2001. The Hf isotope composition of global seawater and the evolution of Hf isotopes in the deep Pacific Ocean from Fe-Mn crusts. Chemical Geology, 178 (1-4), 23-42.

Davis, D. W. and Krogh, T. E., 2001. Preferential dissolution of U-234 and radiogenic Pb from alpha-recoil-damaged lattice sites in zircon: implications for thermal histories and Pb isotopic fractionation in the near surface environment. Chemical Geology, 172 (1-2), 41-58.

de Vernal, A. and Hillaire-Marcel, C., 2000. Sea-ice cover, sea-surface salinity and halo-/thermocline structure of the northwest North Atlantic: modern versus full glacial conditions. Quaternary Science Reviews, 19 (1-5), 65-85.

de Vernal, A., Hillaire-Marcel, C., Peltier, W. R. and Weaver, A. J., 2002. Structure of the upper water column in the northwest North Atlantic: Modern versus last glacial maximum conditions. Paleoceanography, 17 (4), doi:10.1029/2001PA000665.

deVernal, A., Hillaire-Marcel, C. and Bilodeau, G., 1996. Reduced meltwater outflow from the Laurentide ice margin during the Younger Dryas. Nature, 381 (6585), 774-777.

Dickson, R. R. and Brown, J., 1994. The production of North Atlantic Deep Water - sources, rates, and pathways. Journal of Geophysical Research-Oceans, 99 (C6), 12319-12341.

Dosseto, A., Bourdon, B., Gaillardet, J., Allegre, C. J. and Filizola, N., 2006. Time scale and conditions of weathering under tropical climate: Study of the Amazon basin with U-series. Geochimica et Cosmochimica Acta, 70 (1), 71-89.

Dunk, R. M., Mills, R. A. and Jenkins, W. J., 2002. A reevaluation of the oceanic uranium budget for the Holocene. Chemical Geology, 190 (1-4), 45-67.

Duplessy, J. C., Shackleton, N. J., Fairbanks, R. G., Labeyrie, L., Oppo, D. W. and Kallell, N., 1988. Deep-water source variations during the last climatic cycle and their impact on the global deep-water circulation. Paleoceanography, 3, 343-360.

Dyke, A. S. and Prest, V. K., 1987. Late Wisconsinan and Holocene history of the Laurentide ice sheet. Geographie Physique et Quaternaire, 41, 237-264.

Dyke, A. S., Andrews, J. T., Clark, P. U., England, J. H., Miller, G. H., Shaw, J. and Veillette, J. J., 2002. The Laurentide and Innuitian ice sheets during the Last Glacial Maximum. Quaternary Science Reviews, 21 (1-3), 9-31.

Eittreim, S., Ewing, M. and Thorndik.Em, 1969. Suspended matter along continental margin of North American basin. Deep-Sea Research, 16 (6), 613-622.

Eittreim, S., Thorndike, E. M. and Sullivan, L., 1976. Turbidity distribution in Atlantic Ocean. Deep-Sea Research, 23 (12), 1115-1127.

Elderfield, H. and Sholkovitz, E. R., 1987. Rare earth elements in the pore waters of reducing nearshore sediments. Earth and Planetary Science Letters, 82 (3-4), 280-288.

Erel, Y., Harlavan, Y. and Blum, J. D., 1994. Lead isotope systematics of granitoid weathering. Geochimica et Cosmochimica Acta, 58 (23), 5299-5306.

Erel, Y., Blum, J. D., Roueff, E. and Ganor, J., 2004. Lead and strontium isotopes as monitors of experimental granitoid mineral dissolution. Geochimica et Cosmochimica Acta, 68 (22), 4649-4663.

Fagel, N., Innocent, C., Gariepy, C. and Hillaire-Marcel, C., 2002. Sources of Labrador Sea sediments since the last glacial maximum inferred from Nd-Pb isotopes. Geochimica et Cosmochimica Acta, 66(14), 2569-2581.

Fagel, N., Hillaire-Marcel, C., Humblet, M., Brasseur, R., Weis, D. and Stevenson, R., 2004. Nd and Pb isotope signatures of the clay-size fraction of Labrador Sea sediments during the Holocene: Implications for the inception of the modern deep circulation pattern. Paleoceanography, 19 (3).

Fairbanks, R. G., 1989. A 17,000-year glacio-eustatic sea-level record - influence of glacial melting rates on the Younger Dryas event and deep-ocean circulation. Nature, 342 (6250), 637-642.

Falina, A., Sokov, A. and Sarafanov, A., 2007. Variability and renewal of Labrador Sea Water in the Irminger Basin in 1991–2004. Journal of Geophysical Research, 112 Art. No. C01006.

Fisher, T. G., 2007. Abandonment chronology of glacial Lake Agassiz's Northwestern outlet. Palaeogeography, Palaeoclimatology, Palaeoecology, 246 (1), 31-44.

Foster, G. L. and Vance, D., 2006. Negligible glacial-interglacial variation in continental chemical weathering rates. Nature, 444 (7121), 918-921.

Foster, G. L., Vance, D. and Prytulak, J., 2007. No change in the neodymium isotope composition of deep water exported from the North Atlantic on glacial-interglacial timescales. Geology, 35 (1), 37-40.

Francois, R., Frank, M., van der Loeff, M. M. R. and Bacon, M. P., 2004. Th-230 normalization: An essential tool for interpreting sedimentary fluxes during the late Quaternary. Paleoceanography, 19(1), PA1018, doi:10.1029/2003PA000939.

Frank, M., O'Nions, R. K., Hein, J. R. and Banakar, V. K., 1999a. 60 Myr records of major elements and Pb-Nd isotopes from hydrogenous ferromanganese crusts: reconstruction of seawater paleochemistry. Geochimica et Cosmochimica Acta, 63 (11-12), 1689-1708.

Frank, M., Reynolds, B. C. and O'Nions, R. K., 1999b. Nd and Pb isotopes in Atlantic and Pacific water masses before and after closure of the Panama gateway. Geology, 27 (12), 1147-1150.

Frank, M., 2002. Radiogenic isotopes: Tracers of past ocean circulation and erosional input. Reviews of Geophysics, 40 (1), 1001, doi:10.1029/2000 RG000094.

Frank, M., Whiteley, N., Kasten, S., Hein, J. R. and O'Nions, R. K., 2002. North Atlantic Deep Water export to the Southern Ocean over the past 14 Myr: Evidence from Nd and Pb isotopes in ferromanganese crusts. Paleocanography, 17 (2), 1022, doi:10.1029/2000PA000606.

Frank, M., Marbler, H., Koschinsky, A., de Flierdt, T. V., Klemm, V., Gutjahr, M., Halliday, A. N., Kubik, P. W. and Halbach, P., 2006. Submarine hydrothermal venting related to volcanism in the Lesser Antilles: Evidence from ferromanganese precipitates. Geochemistry Geophysics Geosystems, 7.

Galer, S. J. G. and Abouchami, W., 1998. Practical application of lead triple spiking for correction of instrumental mass discrimination. Min. Mag., 62A, 491–492.

Ganopolski, A. and Rahmstorf, S., 2001. Rapid changes of glacial climate simulated in a coupled climate model. Nature, 409 (6817), 153-158.

Gherardi, J. M., Labeyrie, L., McManus, J. F., Francois, R., Skinner, L. C. and Cortijo, E., 2005. Evidence from the Northeastern Atlantic basin for variability in the rate of the meridional overturning circulation through the last deglaciation. Earth and Planetary Science Letters, 240 (3-4), 710-723.

Godfrey, L. V., White, W. M. and Salters, V. J. M., 1996. Dissolved zirconium and hafnium distributions across a shelf break in the northeastern Atlantic Ocean. Geochimica et Cosmochimica Acta, 60 (21), 3995-4006.

Godfrey, L. V., Lee, D. C., Sangrey, W. F., Halliday, A. N., Salters, V. J. M., Hein, J. R. and White, W. M., 1997. The Hf isotopic composition of ferromanganese nodules and crusts and hydrothermal manganese deposits: Implications for seawater Hf. Earth and Planetary Science Letters, 151 (1-2), 91-105.

Goldstein, S. J. and Jacobsen, S. B., 1987. The Nd and Sr isotopic systematics of river-water dissolved material - implications for the sources of Nd and Sr in seawater. Chemical Geology, 66 (3-4), 245-272.

Goldstein, S. L. and Hemming, S. R., 2003. Long-lived isotopic tracers in oceanography, paleoceanography, and ice-sheet dynamics. Treatise on Geochemistry, Vol. 6, 453-489.

Guo, L. D., Santschi, P. H., Baskaran, M. and Zindler, A., 1995. Distribution of dissolved and particulate Th-230 and Th-232 in seawater from the Gulf of Mexico and off Cape Hatteras as measured by Sims. Earth and Planetary Science Letters, 133 (1-2), 117-128.

Gwiazda, R. H., Hemming, S. R. and Broecker, W. S., 1996. Provenance of icebergs during Heinrich event 3 and the contrast to their sources during other Heinrich episodes. Paleoceanography, 11 (4), 371-378.

Haley, B. A., Klinkhammer, G. P. and McManus, J., 2004. Rare earth elements in pore waters of marine sediments. Geochimica et Cosmochimica Acta, 68(6), 1265-1279.

Halliday, A. N., Davidson, J. P., Holden, P., Owen, R. M. and Olivarez, A. M., 1992. Metalliferous sediments and the scavenging residence time of Nd near hydrothermal vents. Geophysical Research Letters, 19 (8), 761-764.

Hannigan, R. E. and Sholkovitz, E. R., 2001. The development of middle rare earth element enrichments in freshwaters: weathering of phosphate minerals. Chemical Geology, 175 (3-4), 495-508.

Hansen, B. T. and Friderichsen, J. D., 1989. The influence of recent lead loss on the interpretation of disturbed U---Pb systems in zircons from igneous rocks in East Greenland. Lithos, 23 (3), 209-223.

Harlavan, Y., Erel, Y. and Blum, J. D., 1998. Systematic Changes in Lead Isotopic Composition with Soil Age in Glacial Granitic Terrains. Geochimica et Cosmochimica Acta, 62 (1), 33-46.

Harlavan, Y. and Erel, Y., 2002. The release of Pb and REE from granitoids by the dissolution of accessory phases. Geochimica et Cosmochimica Acta, 66 (5), 837-848.

Haskell, B. J. and Johnson, T. C., 1993. Surface sediment response to deep-water circulation on the Blake Outer Ridge, Western North Atlantic - paleoceanographic implications. Sedimentary Geology, 82 (1-4), 133-144.

Haskell, B. J. J., T.C., 1991. Fluctuations in deep western North Atlantic circulation on the Blake Outer Ridge during the last deglaciation. Paleoceanography, 6, 21-31.

Hein, J. R., Koschinsky, A., Bau, M., Manheim, F. T., Kang, J.-K. and Roberts, L., 1999. Cobalt-rich ferromanganese crusts in the Pacific. In: D.S. Cronan (Editor), Handbook of marine mineral deposits. CRC Press, pp. 239-279.

Hemming, S. R., 2004. Heinrich events: Massive late pleistocene detritus layers of the North Atlantic and their global climate imprint. Reviews of Geophysics, 42 (1), RG1005, doi:10.1029/2003 RG000128.

Henderson, G. M., Martel, D. J., Onions, R. K. and Shackleton, N. J., 1994. Evolution of seawater Sr-87/Sr-86 over the last 400 ka - the absence of glacial-interglacial cycles. Earth and Planetary Science Letters, 128 (3-4), 643-651.

Henderson, G. M. and Maier-Reimer, E., 2002. Advection and removal of 210Pb and stable Pb isotopes in the oceans: a general circulation model study. Geochimica et Cosmochimica Acta, 66 (2), 257-272.

Hillaire-Marcel, C., de Vernal, A., Bilodeau, G. and Weaver, A. J., 2001. Absence of deep-water formation in the Labrador Sea during the last interglacial period. Nature, 410(6832), 1073-1077.

Horwitz, E. P., Chiarizia, R. and Dietz, M. L., 1992. A novel strontium-selective extraction chromatographic resin. Solvent Extraction and Ion Exchange, 10 (2), 313-336.

Hughen, K. A., Baillie, M. G. L., Bard, E., Beck, J. W., Bertrand, C. J. H., Blackwell, P. G., Buck, C. E., Burr, G. S., Cutler, K. B., Damon, P. E., Edwards, R. L., Fairbanks, R. G., Friedrich, M., Guilderson, T. P., Kromer, B., McCormac, G., Manning, S., Ramsey, C. B., Reimer, P. J., Reimer, R. W., Remmele, S., Southon, J. R., Stuiver, M., Talamo, S., Taylor, F. W., van der Plicht, J. and Weyhenmeyer, C. E., 2004a. Marine04 marine radiocarbon age calibration, 0–26 cal kyr BP. Radiocarbon, 46, 1059-1086.

Hughen, K. A., Southon, J. R., Bertrand, C. J. H., Frantz, B. and Zermeno, P., 2004b. Cariaco Basin calibration update: Revisions to calendar and 14C chronologies for core PL07-58PC. Radiocarbon, 46, 1161-1187.

Hunt, R. E., Swift, D. J. P. and Palmer, H., 1977. Constructional shelf topography, Diamond Shoals, North Carolina. Geological Society of America Bulletin, 88 (2), 299-311.

Jacobsen, S. B. and Wasserburg, G. J., 1980. Sm-Nd isotopic evolution of chondrites. Earth and Planetary Science Letters, 50 (1), 139-155.

Jeandel, C., 1993. Concentration and isotopic composition of Nd in the South Atlantic Ocean. Earth and Planetary Science Letters, 117, 581-591.

Jeandel, C., Bishop, J. K. and Zindler, A., 1995. Exchange of neodymium and its isotopes between seawater and small and large particles in the Sargasso Sea. Geochimica et Cosmochimica Acta, 59 (3), 535-547.

Jennings, A. E., Hald, M., Smith, M. and Andrews, J. T., 2006. Freshwater forcing from the Greenland Ice Sheet during the Younger Dryas: evidence from southeastern Greenland shelf cores. Quaternary Science Reviews, 25 (3-4), 282-298.

Johnsen, S. J., Clausen, H. B., Dansgaard, W., Fuhrer, K., Gundestrup, N., Hammer, C. U., Iversen, P., Jouzel, J., Stauffer, B. and steffensen, J. P., 1992. Irregular glacial interstadials recorded in a new Greenland ice core. 359 (6393), 311-313.

Jones, C. E., Halliday, A. N., Rea, D. K. and Owen, R. M., 2000. Eolian inputs of lead to the North Pacific. Geochimica et Cosmochimica Acta, 64 (8), 1405-1416.

Keigwin, L. D., Jones, G. A., Lehman, S. J. and Boyle, E. A., 1991. Deglacial meltwater discharge, North Atlantic deep circulation, and abrupt climate change. Journal of Geophysical Research-Oceans, 96 (C9), 16811-16826.

Keigwin, L. D., 2004. Radiocarbon and stable isotope constraints on Last Glacial Maximum and Younger Dryas ventilation in the western North Atlantic. Paleoceanography, 19 (4), PA4012, doi:10.1029/2004 PA001029.

Kennett, J. P. and Shackleton, N. J., 1975. Laurentide Ice Sheet meltwater recorded in Gulf of Mexico deep-sea cores. Science, 188 (4184), 147-150.

Klemm, V., Levasseur, S., Frank, M., Hein, J. R. and Halliday, A. N., 2005. Osmium isotope stratigraphy of a marine ferromanganese crust. Earth and Planetary Science Letters, 238 (1-2), 42-48.

Klemm, V., Reynolds, B., Frank, M., Pettke, T. and Halliday, A. N., 2007. Cenozoic changes in atmospheric lead recorded in central Pacific ferromanganese crusts. Earth and Planetary Science Letters, 253 (1-2), 57-66.

Labeyrie, L. D., Duplessy, J. C., Duprat, J., Juilletleclerc, A., Moyes, J., Michel, E., Kallel, N. and Shackleton, N. J., 1992. Changes in the vertical structure of the North Atlantic Ocean between glacial and modern times. Quaternary Science Reviews, 11 (4), 401-413.

Lacan, F. and Jeandel, C., 2004a. Subpolar Mode Water formation traced by neodymium isotopic composition. Geophysical Research Letters, 31 (14), L14306, doi:10.1029/2004GL019747.

Lacan, F. and Jeandel, C., 2004b. Neodymium isotopic composition and rare earth element concentrations in the deep and intermediate Nordic Seas: Constraints on the Iceland Scotland Overflow Water signature. Geochemistry Geophysics Geosystems, 5, Q11006, doi:10.1029/2004GC000742.

Lacan, F. and Jeandel, C., 2005a. Acquisition of the neodymium isotopic composition of the North Atlantic Deep Water. Geochemistry Geophysics Geosystems, 6, Q12008, doi:10.1029/2005GC000956.

Lacan, F. and Jeandel, C., 2005b. Neodymium isotopes as a new tool for quantifying exchange fluxes at the continent-ocean interface. Earth and Planetary Science Letters, 232 (3-4), 245-257.

Lee, D. C., Halliday, A. N., Christensen, J. N., Burton, K. W., Hein, J. R. and Godfrey, L. V., 1998. High resolution Hf isotope stratigraphy of Fe-Mn crusts. American Geophysical Union Fall Meeting conference abstract.

Lee, D. C., Halliday, A. N., Hein, J. R., Burton, K. W., Christensen, J. N. and Gunther, D., 1999. Hafnium isotope stratigraphy of ferromanganese crusts. Science, 285 (5430), 1052-1054.

Levasseur, S., Birck, J.-L. and Allègre, C. J., 1998. Direct measurement of femtomoles of osmium and the $^{187}Os/^{186}Os$ ratio in seawater. Science, 282, 272-274.

Leventer, A., Williams, D. F. and Kennett, J. P., 1982. Dynamics of the Laurentide ice sheet during the last deglaciation: evidence from the Gulf of Mexico. Earth and Planetary Science Letters, 59 (1), 11-17.

Licciardi, J. M., Clark, P. U., Jenson, J. W. and Macayeal, D. R., 1998. Deglaciation of a soft-bedded Laurentide ice sheet. Quaternary Science Reviews, 17 (4-5), 427-448.

Llave, E., Schonfeld, J., Hernandez-Molina, F. J., Mulder, T., Somoza, L., Diaz del Rio, V. and Sanchez-Almazo, I., 2006. High-resolution stratigraphy of the Mediterranean outflow contourite system in the Gulf of Cadiz during the late Pleistocene: The impact of Heinrich events. Marine Geology, 227 (3-4), 241-262.

Lugmair, G. W. and Galer, S. J. G., 1992. Age and isotopic relationships among the angrites Lewis Cliff 86010 and Angra dos Reis. Geochimica et Cosmochimica Acta, 56 (4), 1673-1694.

Luo, S. D., Ku, T. L., Wang, L., Southon, J. R., Lund, S. P. and Schwartz, M., 2001. Al-26, Be-10 and U-Th isotopes in Blake Outer Ridge sediments: implications for past changes in boundary scavenging. Earth and Planetary Science Letters, 185 (1-2), 135-147.

Luo, X. Z., Rehkamper, M., Lee, D. C. and Halliday, A. N., 1997. High precision Th-230/Th-232 and U-234/U-238 measurements using energy-filtered ICP magnetic sector multiple collector mass spectrometry. International Journal of Mass Spectrometry, 171 (1-3), 105-117.

Lynch-Stieglitz, J., Curry, W. B. and Slowey, N., 1999. Weaker Gulf Stream in the Florida straits during the last glacial maximum. Nature, 402 (6762), 644-648.

Marchitto, T. M., Oppo, D. W. and Curry, W. B., 2002. Paired benthic foraminiferal Cd/Ca and Zn/Ca evidence for a greatly increased presence of Southern Ocean Water in the glacial North Atlantic. Paleoceanography, 17, doi:10.1029/2000PA000598.

Marchitto, T. M. and Broecker, W. S., 2006. Deep water mass geometry in the glacial Atlantic Ocean: A review of constraints from the paleonutrient proxy Cd/Ca. Geochemistry Geophysics Geosystems, 7.

Marshall, S. J. and Clarke, G. K. C., 1999. Modeling North American freshwater runoff through the last glacial cycle. Quaternary Research, 52 (3), 300-315.

McCartney, M. S. and Talley, L. D., 1982. The sub-polar mode water of the North Atlantic ocean. Journal of Physical Oceanography, 12 (11), 1169-1188.

McCave, I. N., 1986. Local and global aspects of the bottom nepheloid layers in the world ocean. Netherlands Journal of Sea Research, 20 (2-3), 167-181.

McKelvey, B. A. and Orians, K. J., 1998. The determination of dissolved zirconium and hafnium from seawater using isotope dilution inductively coupled plasma mass spectrometry. Marine Chemistry, 60 (3-4), 245-255.

McLennan, S. M., 1989. Rare earth elements in sedimentary rocks - influence of provenance and sedimentary processes. Reviews in Mineralogy, 21, 169-200.

McManus, J. F., Francojs, R., Gherardi, J. M., Keigwin, L. D. and Brown-Leger, S., 2004. Collapse and rapid resumption of Atlantic meridional circulation linked to deglacial climate changes. Nature, 428 (6985), 834-837.

Münker, C., Weyer, S., Scherer, E. and Mezger, K., 2001. Separation of high field strength elements (Nb, Ta, Zr, Hf) and Lu from rock samples for MC-ICPMS measurements. Geochemistry Geophysics Geosystems, 2, art. no.-2001GC000183.

Murakami, T., Chakoumakos, B. C., Ewing, R. C., Lumpkin, G. R. and Weber, W. J., 1991. Alpha-decay event damage in zircon. Am. Mineral., 76 (9-10), 1510-1532.

Nesbitt, H. W. and Young, G. M., 1982. Early Proterozoic climates and plate motions inferred from major element chemistry of lutites. Nature, 299 (5885), 715-717.

Nesbitt, H. W. and Young, G. M., 1984. Prediction of some weathering trends of plutonic and volcanic rocks based on thermodynamic and kinetic considerations. Geochimica et Cosmochimica Acta, 48 (7), 1523-1534.

Nier, A. O., 1938. The isotopic constitution of strontium, barium, bismuth, thallium and mercury. Physical Review, 54 (4), 275-278.

Nowell, G. M., Kempton, P. D., Noble, S. R., Fitton, J. G., Saunders, A. D., Mahoney, J. J. and Taylor, R. N., 1998. High precision Hf isotope measurements of MORB and OIB by thermal ionisation mass spectrometry: insights into the depleted mantle. Chemical Geology, 149, 211-233.

Nozaki, Y., 1991. The systematics and kinetics of U-Th decay series nuclides in ocean water. Reviews in Aquatic Sciences, 4 (1), 75-105.

O'Nions, R. K., Carter, S. R., Cohen, R. S., Evensen, N. M. and Hamilton, P. J., 1978. Pb, Nd and Sr isotopes in oceanic ferromanganese deposits and ocean-floor basalts. Nature, 273 (5662), 435-438.

Ohr, M., Halliday, A. N. and Peacor, D. R., 1994. Mobility and fractionation of rare earth elements in argillaceous sediments: Implications for dating diagenesis and low-grade metamorphism. Geochimica et Cosmochimica Acta, 58 (1), 289-312.

Palmer, M. R. and Elderfield, H., 1985a. Sr isotope composition of sea water over the past 75 Myr. Nature, 314 (6011), 526-528.

Palmer, M. R. and Elderfield, H., 1985b. Variations in the Nd isotopic composition of foraminifera from Atlantic Ocean sediments. Earth and Planetary Science Letters, 73 (2-4), 299-305.

Patchett, P. J., White, W. M., Feldmann, H., Kielinczuk, S. and Hofmann, A. W., 1984. Hafnium rare-earth element fractionation in the sedimentary system and crustal recycling into the Earths mantle. Earth and Planetary Science Letters, 69 (2), 365-378.

Petit, J. R., Jouzel, J., Raynaud, D., Barkov, N. I., Barnola, J. M., Basile, I., Bender, M., Chappellaz, J., Davis, M., Delaygue, G., Delmotte, M., Kotlyakov, V. M., Legrand, M., Lipenkov, V. Y., Lorius, C., Pepin, L., Ritz, C., Saltzman, E. and Stievenard, M., 1999. Climate and atmospheric history of the past 420,000 years from the Vostok ice core, Antarctica. Nature, 399 (6735), 429-436.

Peucker-Ehrenbrink, B. and Ravizza, G., 2000. The marine osmium isotope record. Terra Nova, 12, 205-219.

Pickart, R. S., Straneo, F. and Moore, G. W. K., 2003. Is Labrador Sea Water formed in the Irminger basin? Deep-Sea Research Part I-Oceanographic Research Papers, 50 (1), 23-52.

Piepgras, D. J. and Wasserburg, G. J., 1982. Isotopic composition of neodymium in waters from the Drake Passage. Science, 217 (4556), 207-214.

Piepgras, D. J. and Wasserburg, G. J., 1987. Rare earth element transport in the western North Atlantic inferred from Nd isotopic observations. Geochimica et Cosmochimica Acta, 51 (5), 1257-1271.

Piotrowski, A. M., Lee, D.-C., Christensen, J. N., Burton, K. W., Halliday, A. N., Hein, J. R. and Günther, D., 2000. Changes in erosion and ocean circulation recorded in the Hf isotopic compositions of North Atlantic and Indian Ocean ferromanganese crusts. Earth and Planetary Science Letters, 181 (3), 315-325.

Piotrowski, A. M., 2004. A high resolution record of ocean circulation during the Last Glacial Cycle from Neodymium isotopes. Thesis - Columbia University.

Piotrowski, A. M., Goldstein, S. L., Hemming, S. R. and Fairbanks, R. G., 2004. Intensification and variability of ocean thermohaline circulation through the last deglaciation. Earth and Planetary Science Letters, 225 (1-2), 205-220.

Piotrowski, A. M., Goldstein, S. L., Hemming, S. R. and Fairbanks, R. G., 2005. Temporal relationships of carbon cycling and ocean circulation at glacial boundaries. Science, 307 (5717), 1933-1938.

Piper, D. J. W., Shaw, J. and Skene, K. I., 2007. Stratigraphic and sedimentological evidence for late Wisconsinan sub-glacial outburst floods to Laurentian Fan. Palaeogeography, Palaeoclimatology, Palaeoecology, 246 (1), 101-119.

Rasmussen, T. L., Oppo, D. W., Thomsen, E. and Lehman, S. J., 2003. Deep sea records from the southeast Labrador Sea: Ocean circulation changes and ice-rafting events during the last 160,000 years. Paleoceanography, 18(1), 1018, doi:10.1029/2001PA000736.

Rayburn, J. A., Franzi, D. A. and Knuepfer, P. L. K., 2007. Evidence from the Lake Champlain Valley for a later onset of the Champlain Sea and implications for late glacial meltwater routing to the North Atlantic. Palaeogeography, Palaeoclimatology, Palaeoecology 246 (1), 62-74.

Raymo, M. E. and Ruddiman, W. F., 1992. Tectonic forcing of late Cenozoic climate. Nature, 359, 117-122.

Rehkamper, M., Schonbachler, M. and Stirling, C. H., 2001. Multiple collector ICP-MS: Introduction to instrumentation, measurement techniques and analytical capabilities. Geostandards Newsletter-the Journal of Geostandards and Geoanalysis, 25 (1), 23-40.

Reverdin, G., Niiler, P. P. and Valdimarsson, H., 2003. North Atlantic Ocean surface currents. Journal of Geophysical Research-Oceans, 108 (C1), 3002, doi:10.1029/2001JC001020.

Reynolds, B. C., Frank, M. and O'Nions, R. K., 1999. Nd- and Pb-isotope time series from Atlantic ferromanganese crusts: implications for changes in provenance and paleocirculation over the last 8 Myr. Earth and Planetary Science Letters, 173 (4), 381-396.

Rickaby, R. E. M. and Elderfield, H., 2005. Evidence from the high-latitude North Atlantic for variations in Antarctic Intermediate Water flow during the last deglaciation. Geochemistry Geophysics Geosystems, 6 Art. No. Q05001.

Robinson, L. F., Belshaw, N. S. and Henderson, G. M., 2004. U and Th concentrations and isotope ratios in modern carbonates and waters from the Bahamas. Geochimica et Cosmochimica Acta, 68 (8), 1777-1789.

Robinson, L. F., Adkins, J. F., Keigwin, L. D., Southon, J., Fernandez, D. P., Wang, S. L. and Scheirer, D. S., 2005. Radiocarbon variability in the western North Atlantic during the last deglaciation. Science, 310 (5753), 1469-1473.

Romer, R. L., 2003. Alpha-recoil in U-Pb geochronology: effective sample size matters. Contributions to Mineralogy and Petrology, 145 (4), 481-491.

Rutberg, R. L., Hemming, S. R. and Goldstein, S. L., 2000. Reduced North Atlantic Deep Water flux to the glacial Southern Ocean inferred from neodymium isotope ratios. Nature, 405 (6789), 935-938.

Ruttenberg, K. C., 1992. Development of a sequential extraction method for different forms of phosphorus in marine sediments. Limnology and Oceanography, 37 (7), 1460-1482.

Sarnthein, M., Winn, K., Jung, S. J. A., Duplessy, J. C., Labeyrie, L., Erlenkeuser, H. and Ganssen, G., 1994. Changes in East Atlantic Deep-Water Circulation over the Last 30,000 Years - 8 Time Slice Reconstructions. Paleoceanography, 9, 209-267.

Schaefer, J. M., Denton, G. H., Barrell, D. J. A., Ivy-Ochs, S., Kubik, P. W., Andersen, B. G., Phillips, F. M., Lowell, T. V. and Schluchter, C., 2006. Near-synchronous interhemispheric termination of the Last Glacial Maximum in mid-latitudes. Science, 312 (5779), 1510-1513.

Schaule, B. K. and Patterson, C. C., 1981. Lead concentrations in the Northeast Pacific - Evidence for global anthropogenic perturbations. Earth and Planetary Science Letters, 54 (1), 97-116.

Scher, H. D. and Martin, E. E., 2004. Circulation in the Southern Ocean during the Paleogene inferred from neodymium isotopes. Earth and Planetary Science Letters, 228 (3-4), 391-405.

Schmitz, W. J. and McCartney, M. S., 1993. On the North Atlantic circulation. Reviews of Geophysics, 31 (1), 29-49.

Schmitz, W. J., 1995. On the interbasin-scale thermohaline circulation. Reviews of Geophysics, 33 (2), 151-173.

Schroder-Ritzrau, A., Mangini, A. and Lomitschka, M., 2003. Deep-sea corals evidence periodic reduced ventilation in the North Atlantic during the LGM/Holocene transition. Earth and Planetary Science Letters, 216 (3), 399-410.

Shackleton, N. J., Backman, J., Zimmerman, H., Kent, D. V., Hall, M. A., Roberts, D. G., Schnitker, D., Baldauf, J. G., Desprairies, A., Homrighausen, R., Huddlestun, P., Keene, J. B., Kaltenback, A. J., Krumsiek, K. A. O., Morton, A. C., Murray, J. W. and Westbergsmith, J., 1984. Oxygen isotope calibration of the onset of ice-rafting and history of glaciation in the North Atlantic region. Nature, 307 (5952), 620-623.

Shackleton, N. J., 1987. Oxygen isotopes, ice volume and sea-kevel. Quaternary Science Reviews, 6 (3-4), 183-190.

Speer, K. G. and McCartney, M. S., 1992. Bottom water circulation in the western North Atlantic. Journal of Physical Oceanography, 22 (1), 83-92.

Stahr, F. R. and Sanford, T. B., 1999. Transport and bottom boundary layer observations of the North Atlantic Deep Western Boundary Current at the Blake Outer Ridge. Deep-Sea Research Part II-Topical Studies in Oceanography, 46 (1-2), 205-243.

Steiger, R. H. and Jager, E., 1977. Subcommission on geochronology: Convention on the use of decay constants in geo- and cosmochronology. Earth and Planetary Science Letters, 36 (3), 359-362.

Stuiver, M., Reimer, P. J., Bard, E., Beck, J. W., Burr, G. S., Hughen, K. A., Kromer, B., McCormac, G., Van der Plicht, J. and Spurk, M., 1998. INTCAL98 radiocarbon age calibration, 24,000-0 cal BP. Radiocarbon, 40 (3), 1041-1083.

Suman, D. O. and Bacon, M. P., 1989. Variations in Holocene sedimentation in the North American basin determined from Th-230 measurements. Deep-Sea Research Part a-Oceanographic Research Papers, 36, 869-878.

Swarzenski, P. W., Porcelli, D., Andersson, P. S. and Smoak, J. M., 2003. The behavior of U- and Th-series nuclides in the estuarine environment, Uranium-Series Geochemistry. Reviews in Mineralogy & Geochemistry, pp. 577-606.

Tachikawa, K., Jeandel, C. and Roy-Barman, M., 1999. A new approach to the Nd residence time in the ocean: the role of atmospheric inputs. Earth and Planetary Science Letters, 170 (4), 433-446.

Tachikawa, K., Roy-Barman, M., Michard, A., Thouron, D., Yeghicheyan, D. and Jeandel, C., 2004. Neodymium isotopes in the Mediterranean Sea: Comparison between seawater and sediment signals. Geochimica et Cosmochimica Acta, 68 (14), 3095-3106.

Talley, L. D. and McCartney, M. S., 1982. Distribution and circulation of Labrador Sea Water. Journal of Physical Oceanography, 12 (11), 1189-1205.

Tanaka, T., Togashi, S., Kamioka, H., Amakawa, H., Kagami, H., Hamamoto, T., Yuhara, M., Orihashi, Y., Yoneda, S., Shimizu, H., Kunimaru, T., Takahashi, K., Yanagi, T., Nakano, T., Fujimaki, H., Shinjo, R., Asahara, Y., Tanimizu, M. and

175

Dragusanu, C., 2000. JNdi-1: a neodymium isotopic reference in consistency with LaJolla neodymium. Chemical Geology, 168 (3-4), 279-281.

Tarasov, L. and Peltier, W. R., 2005. Arctic freshwater forcing of the Younger Dryas cold reversal. Nature, 435 (7042), 662-665.

Teller, J. T., Leverington, D. W. and Mann, J. D., 2002. Freshwater outbursts to the oceans from glacial Lake Agassiz and their role in climate change during the last deglaciation. Quaternary Science Reviews, 21 (8-9), 879-887.

Tessier, A., Campbell, P. G. C. and Bisson, M., 1979. Sequential extraction procedure for the speciation of particulate trace metals. Analytical Chemistry, 51 (7), 844-851.

Thirlwall, M. F., 2002. Multicollector ICP-MS analysis of Pb isotopes using a ^{207}Pb-^{204}Pb double spike demonstrates up to 400 ppm/amu systematic errors in Tl-normalization. Chemical Geology, 184 (3-4), 255-279.

Thomson, J., Colley, S., Anderson, R., Cook, G. T., Mackenzie, A. B. and Harkness, D. D., 1993a. Holocene Sediment Fluxes in the Northeast Atlantic from Th-230(Excess) and Radiocarbon Measurements. Paleoceanography, 8, 631-650.

Thomson, J., Higgs, N. C., Croudace, I. W., Colley, S. and Hydes, D. J., 1993b. Redox zonation of elements at an oxic/post-oxic boundary in deep-sea sediments. Geochimica et Cosmochimica Acta, 57 (3), 579-595.

Thomson, J., Higgs, N. C., Wilson, T. R. S., Croudace, I. W., De Lange, G. J. and Van Santvoort, P. J. M., 1995. Redistribution and geochemical behaviour of redox-sensitive elements around S1, the most recent eastern Mediterranean sapropel. Geochimica et Cosmochimica Acta, 59 (17), 3487-3501.

Tovar-Sanchez, A., Sanudo-Wilhelmy, S. A., Garcia-Vargas, M., Weaver, R. S., Popels, L. C. and Hutchins, D. A., 2003. A trace metal clean reagent to remove surface-bound iron from marine phytoplankton. Marine Chemistry, 82 (1-2), 91-99.

van de Flierdt, T., Frank, M., Lee, D.-C. and Halliday, A. N., 2002. Glacial weathering and the hafnium isotope composition of seawater. Earth and Planetary Science Letters, 198, 167-175.

van de Flierdt, T., Frank, M., Halliday, A. N., Hein, J. R., Hattendorf, B., Gunther, D. and Kubik, P. W., 2003. Lead isotopes in North Pacific deep water - implications for past changes in input sources and circulation patterns. Earth and Planetary Science Letters, 209 (1-2), 149-164.

van de Flierdt, T., Frank, M., Halliday, A. N., Hein, J. R., Hattendorf, B., Gunther, D. and Kubik, P. W., 2004a. Deep and bottom water export from the Southern Ocean to the Pacific over the past 38 million years. Paleoceanography, 19 (1).

van de Flierdt, T., Frank, M., Lee, D. C., Halliday, A. N., Reynolds, B. C. and Hein, J. R., 2004b. New constraints on the sources and behavior of neodymium and hafnium in seawater from Pacific Ocean ferromanganese crusts. Geochimica et Cosmochimica Acta, 68 (19), 3827-3843.

van de Flierdt, T. V., Frank, M., Halliday, A. N., Hein, J. R., Hattendorf, B., Gunther, D. and Kubik, P. W., 2004c. Tracing the history of submarine hydrothermal inputs and the significance of hydrothermal hafnium for the seawater budget-a combined Pb-Hf-Nd isotope approach. Earth and Planetary Science Letters, 222 (1), 259-273.

van de Flierdt, T. R., L.F.; Adkins, J.F.; Hemming, S.R.; Goldstein, S.J., 2006. Temporal stability of the neodymium isotope signature of the Holocene to glacial North Atlantic. Paleoceanography, 21 Art. No. PA4102.

Vance, D. and Burton, K., 1999. Neodymium isotopes in planktonic foraminifera: a record of the response of continental weathering and ocean circulation rates to climate change. Earth and Planetary Science Letters, 173 (4), 365-379.

Vautravers, M. J., Shackleton, N. J., Lopez-Martinez, C. and Grimalt, J. O., 2004. Gulf Stream variability during marine isotope stage 3. Paleoceanography, 19 (2).

Vervoort, J. D., Patchett, P. J., Blichert-Toft, J. and Albarède, F., 1999. Relationship between Lu-Hf and Sm-Nd isotopic systems in the global sedimentary system. Earth and Planetary Science Letters, 168, 79-99.

Volkening, J., Walczyk, T. and Heumann, K. G., 1991. Osmium isotope ratio determinations by negative thermal ionization mass-spectrometry. International Journal of Mass Spectrometry and Ion Processes, 105 (2), 147-159.

von Blanckenburg, F., Onions, R. K. and Hein, J. R., 1996. Distribution and sources of pre-anthropogenic lead isotopes in deep ocean water from Fe-Mn crusts. Geochimica Et Cosmochimica Acta, 60 (24), 4957-4963.

von Blanckenburg, F. and Nagler, T. F., 2001. Weathering versus circulation-controlled changes in radiogenic isotope tracer composition of the Labrador Sea and North Atlantic Deep Water. Paleoceanography, 16 (4), 424-434.

Walder, A. J. and Furuta, N., 1993. High-precision lead-isotope ratio measurement by inductively-coupled plasma multiple collector mass-spectrometry. Analytical Sciences, 9 (5), 675-680.

White, W. M., Patchett, P. J. and BenOthman, D., 1986. Hf isotope ratios of marine sediments and Mn nodules: evidence for a mantle source of Hf in seawater. Earth and Planetary Science Letters, 79, 46-54.

Wingham, D. J., Siegert, M. J., Shepherd, A. and Muir, A. S., 2006. Rapid discharge connects Antarctic subglacial lakes. Nature, 440 (7087), 1033-1036.

Yokoyama, Y., Lambeck, K., De Deckker, P., Johnston, P. and Fifield, L. K., 2000. Timing of the Last Glacial Maximum from observed sea-level minima. Nature, 406(6797), 713-716.

Zachos, J., Pagani, M., Sloan, L., Thomas, E. and Billups, K., 2001. Trends, rhythms, and aberrations in global climate 65 Ma to present. Science, 292 (5517), 686-693.

Zimmermann, B., Lee, D. C., Porcelli, D., Frank, M., Halliday, A. N., Andersson, P. S. and Baskaran, M., 2004. The Isotopic Composition of Hafnium in Seawater: First Results From the Arctic Ocean. Ocean Sciences Meeting Conference Abstract.

Zimmermann, B. E., Frank, M., Porcelli, D., Lee, D.-C., Halliday, A. N., Andersson, P. S. and Baskaran, M., 2005. Hafnium and neodymium isotope measurements in seawater - a powerful combination of geochemical tracers. EGU Conference abstract.

Appendix

Age Calibration

Published conventional ^{14}C ages of Keigwin (2004) were transformed into calendar years using the marine radiocarbon age calibration Marine04 of Hughen et al. (2004) assuming $\Delta R = 0$ (Table A2). The Pb isotopic radiogenic peak observed in 12JPC– 55cm has been age-matched to the radiogenic Pb isotopic peak seen in 51GGC– 290cm (i.e., 11.2 ka BP), hence supplying a third age tie point for core 12JPC. This age matching is valid because both depths have been ventilated by NADW after the Younger Dryas (see chapter 3). Calibrated ages of other depths in core 12JPC and 51GGC have been linearly interpolated between absolute age tie points and are displayed in Table 4.1. Core 51GGC contains one hiatus between 380 cm and 370 cm, marked by a sharp change in sediment composition. The upper limit of a non-turbiditic sand layer is overlain by silty sediment.

Table A1. Radiocarbon and calibrated ages for cores 51GGC and 12JPC

KNR140 - 51GGC, 1790 m

depth in core (cm)	conventional radiocarbon age	2σ error	calibrated age (calendar years BP)	(- 2σ)	(+ 2σ)	min. age	max. age
280	10,100	110	11,130	90	70	11,040	11,210
290	10,200	110	11,200	70	60	11,120	11,260
300	10,700	120	12,060	130	270	11,930	12,330
310	11,350	130	12,890	50	90	12,840	12,980
312	11,500	150	13,010	120	160	12,890	13,170
320	11,950	120	13,370	110	120	13,260	13,490
330	12,250	200	13,720	270	190	13,460	13,920
350	13,150	170	15,050	320	260	14,730	15,310
360	14,300	170	16,550	380	380	16,170	16,930
370	14,750	160	17,160	410	430	16,750	17,590
380	17,850	220	20,570	310	480	20,260	21,050
400	20,200	260	23,730	430	390	23,290	24,120

KNR140 - 12JPC, 4250 m

depth in core (cm)	conventional radiocarbon age	2σ error	calibrated age (calendar years BP)	(- 2σ)	(+ 2σ)	min. age	max. age
79	10,950	110	12,420	90	340	12,340	12,770
231	15,750	150	18,730	140	110	18,590	18,840

www.ingramcontent.com/pod-product-compliance
Lightning Source LLC
Chambersburg PA
CBHW021051210326
41598CB00016B/1175